1 Bioorganic Chemistry Frontiers

Bioorganic Chemistry Frontiers

Volume 1

Editor-in-Chief: H. Dugas

With contributions by
S.A. Benner, F. Ebmeyer, L. Echegoyen,
A.D. Ellington, T.M. Fyles, G.W. Gokel,
S. Shinkai, F. Vögtle

With 71 Figures and 10 Tables

Springer-Verlag Berlin Heidelberg NewYork
London Paris Tokyo Hong Kong

Editor-in-Chief

Professor H. Dugas
Université de Montréal
Département de Chimie,
Montréal, Quebec H3C 3J7, Canada

ISBN-13: 978-3-642-75258-2 e-ISBN-13: 978-3-642-75256-8
DOI: 10.1007/978-3-642-75256-8

Library of Congress Cataloging-in-Publication Data
Bioorganic chemistry frontiers/editor, H. Dugas; with contributions by
S.A. Benner . . . [et al.]. p. cm.
Includes bibliographical references.
ISBN-13: 978-3-642-75258-2
1. Bioorganic chemistry. I. Dugas, Hermann. II. Benner, Steven A.
(Steven Albert), 1954–
QP550.B56 1990 574. 192-dc20 89-26212 CIP

Softcover reprint of the hardcover 1st edition 1990

Typesetting: Macmillan India Ltd, Bangalore 25; Offset printing: Mercedes-Druck, Berlin

2151/3020-543210 – Printed on acid-free paper

Preface

It is well accepted that progress in biological and biochemical research is based mainly on a better understanding of life processes on a molecular level. For this, modern chemical techniques for structural elucidation even of sophisticated biomolecules and theoretical and mechanistic considerations involving biological macromolecules help us to understand structure–function relations, metabolic processes, molecular and cellular recognition, and the reproduction of life.

On the other hand, controlled manipulation of the structure of biological macromolecules and the synthesis of well designed biomimetic models are the basic tools used in bioorganic chemistry, a field on the border line between classical biochemistry and classical organic chemistry. For this, increasing numbers of chemists and biochemists are studying simple synthetic molecules as models of enzyme action, ion transport across membranes and in general receptor–substrate interaction.

This new series, *Bioorganic Chemistry Frontiers*, will attempt to bring together critical reviews on the progress in this field. In this first volume of the series, five different active domains are covered and are presented to stress the diversity and scope of bioorganic chemistry. The first chapter (Benner and Ellington) addresses the question of protein structure and evolution. Chapters 2 (Fyles) and 3 (Gokel and Echegoyen) focus on ion transport across membranes using crown ethers as biomimetic models. The last two chapters (Ebmeyer and Vögtle, and Shinkai) exploit further the concept of molecular recognition and molecular architecture by designing specific macrocyclic organic receptors or cavities.

This new series should not only fill a current need of covering pinpoint areas at the frontier between biology and chemistry but should also, as a bonus, provide an incentive for further research in bioorganic chemistry.

Montreal, January 1990 H. Dugas

Editorial Board

Table of Contents

Evolution and Structural Theory: The Frontier Between Chemistry and Biology

Steven A. Benner and Andrew D. Ellington
Laboratory for Organic Chemistry, Swiss Federal Institute of Technology, CH-8092 Zurich, Switzerland

Experimental data, hypotheses, and ideas are developed for unifying structural theory from chemistry and evolutionary theory from biology into a coherent package useful for understanding and manipulating biological macromolecules. The discussion focuses on natural selection, neutral drift, and conserved historical accidents as determinants of the behavior of modern proteins. Models are reviewed that interrelate the primary structures of enzymes and their behaviors, primary structure and natural selection, and finally natural selection and enzymic behavior. These models can be applied to guide experiments in protein engineering, predict the tertiary structure of proteins from sequence data, and describe forms of now-extinct life that were present in earlier episodes of natural history, and several illustrative examples of each are given.

1 Introduction

Although much progress has been made over the past three decades in exploring the frontier between chemistry and biology, these two disciplines still cannot be said to have been bridged [1]. This is especially true when one considers the theories central to the two fields, Evolutionary Theory in biology and Structural Theory in organic chemistry. Much work remains to be done to unify these two theories into a coherent picture that can suggest research problems and guide experimentation.

Unifying these theories is difficult because progress comes only by a close interaction between disciplines that traditionally have been widely separated. Such an interaction is essentially impossible in a collaboration between different research groups; biologists and chemists speak different languages, view different research problems as important, and hold firmly to different scientific traditions. However, a single laboratory tackling this problem must have an uncommon experimental scope, and include synthetic chemists, enzymologists and molecular biologists, a scope difficult to manage even under the best of circumstances.

Further, and more seriously, chemists and others with a command of structural theory generally consider biological theory to be an inappropriate focus for their intellectual efforts. This is true even among chemists who believe that biological molecules are interesting research targets. Evolutionary theory is generally viewed as lacking discipline, intellectually "soft", and insufficiently detailed about specific outcomes at the chemical level to be useful to a "rigorous" chemist designing "serious" experiments in a "real" laboratory.

These opinions notwithstanding, there is no inherent obstacle to a "grand unification" of evolutionary and structural theories as they pertain to biological macromolecules. First, the three experimental tools necessary for applying structural theory to a problem (synthesis, purification to constitutional homogeneity, structural analysis at atomic resolution) are now available for biological macromolecules. Further, molecular biological techniques permit direct tests of evolutionary hypotheses. Mutations can be deliberately introduced into genes for proteins, the impact of mutations on the behavior of a protein in vitro can be measured, organisms can be constructed that differ in only by a single mutation, and the survival of such organisms can be measured.

But unification is especially timely because of the explosion in structural information available for biological macromolecules. More than 10 new sequences of proteins appear every week, and this rate has been doubling every other year. Sequence data are accessible through a variety of data bases, although the volume of data has made it all but impossible for these data base services to be kept current [2,3].

Sequence data are constitutional formulas, and structural theory holds that all of the properties of a molecule can be deduced from these formulas alone. The constitutional information found in sequences is supplemented by a collection of conformational information in crystallographic data bases. These data bases are growing, although more slowly than those containing sequence data. This slower growth is less problematical than it might seem. Chothia and Lesk have shown convincingly that the folded form in a protein diverges far more slowly than sequence [4], and models for a protein that has not been studied crystallographically can often be extrapolated from the known structure of a homologous protein [5]. Thus, even if a crystal structure is not available for the particular sequence in hand, if the protein can be shown to be homologous with another protein with known crystal structure, a medium resolution model can usually be obtained.

This review describes progress that has been made in recent years in developing an evolutionary picture of macromolecular chemistry (or a chemical picture of macromolecular evolution). We also offer some new hypotheses and approaches, and hope to stimulate the reader to do the same by pointing out some of the outstanding problems in this often difficult area.

1.1 Describing Macromolecular Behavior Chemically

The constitutional and conformational information contained in the newly available sequence data can most obviously be combined with crystallographic information and molecular biological tools that allow the deliberate replacement of single amino acids in a protein. By replacing individual amino acids, a bioorganic chemist can (at least in principle) determine which residues are responsible for which behaviors in a protein. Such an approach is analogous to that used by physical organic chemists earlier in this century to unravel relationships between structure and function in smaller organic molecules.

In practice, interpreting the impact of point mutations on enzymatic behavior in terms of structure is every bit as problematic in proteins as it is in small molecules. Thus, much "enzyme engineering" is perhaps better characterized as "tinkering" with enzymes [6]. There have been many disappointing results, and much work clearly must be done before the approach yields the insights promised by many [7]. Still, in some enzymes, enough mutants have been made and studied that a comprehensive picture of enzymatic catalysis becomes possible. One of the best examples is tyrosine aminoacyl t-RNA synthetase. A richly documented model for the interaction between this enzyme and its substrate and transition state has emerged from careful kinetic studies of several dozen mutants by Fersht and his coworkers, and it is not inappropriate to say that we understand the chemical basis for catalysis in this enzyme [8].

This is a major accomplishment, as the standard lament of bioorganic chemists is that "very little is known quantitatively about the enormous rate accelerations brought about by enzymes" [9]. This is no longer true. Likewise, it is no longer necessary to argue over broad explanations of enzymic catalysis in terms of "transition state stabilization" or "utilization of intrinsic binding energy" [10]. Such theories were, in any case, little more than thermodynamic tautologies, useful neither for explaining catalysis by enzymes nor designing experiments [11].[1] However, it is clear from Fersht's work that enzyme catalysis can be quantitatively accounted for by the sum of many small differential interactions between the enzyme, the substrate, and the transition state. This statement could have been made from first principles decades ago, but the documentation provided by Fersht (and now by others) provides an accountant's understanding of this statement. It remains only to ask how detailed must the accounting be before one acknowledges that we "understand" catalysis in any individual enzyme, and how many accounting ledgers must be prepared for other enzymes before the task becomes uninteresting.

One risk in this approach is that it might never reveal profound understanding. One can say with absolute confidence that the free energies of interaction between enzyme and transition state, minus the free energies of interaction between enzyme and substrate, will equal exactly the free energy associated with catalysis in every enzyme. One can say with only slightly less confidence that the interactions will be chemically normal, a collection of hydrogen bonds, dipolar interactions, hydrophobic effects, and so on. Each enzyme surely will be different in its details, but will the different details be interesting?

This question can be asked about bioorganic data in general. Bioorganic chemists spend lifetimes collecting data describing differences and similarities between enzymes; kinetics, stereospecificity, substrate specificity, dependence on cofactor, thermal stability, and dynamic behavior are only a few of the properties that are studied. These collections strike the outsider as compilations of trivia. Advanced technology makes these measurements possible. But are these measurements worthwhile?

1.2 Evolution

To address questions at this level, one must turn to biology, and the evolutionary theory that unifies it. Enzymes are not the products of design, but rather arose by a combination of random mutation and natural selection. Thus, not all enzymatic behaviors are rational. At some levels, behavior is adaptive (and therefore understandable in terms of function). At other levels, behavior reflects

[1] Within the context of transition state theory and the laws of thermodynamics, the statement "enzymes bind transition states more tightly than they bind ground states" is logically equivalent to the statement that enzymes catalyze reactions. It is no more an "explanation" for catalysis than the statement "A has a lower free energy than B" is an explanation for the fact that A predominates at equilibrium.

"Intrinsic binding energy" is more problematic. It is not an experimental concept, being originally defined as "the standard free energy change that may be obtained in an ideal situation from the binding forces beween a compound or a substituent group and a macromolecule, in the absence of destabilization or loss of translational, rotational, and internal entropy. It will never be observed. . . . " Even as a theoretical concept, it is compromised. The forces stabilizing an interaction between two molecules are always infinite in the absence of forces destabilizing this interaction; therefore, all intrinsic binding energies are infinite. The problems involved in separating positive and negative interaction are compounded if one attempts to further partition interaction energies among substituents within a single molecule.

conservation (and therefore the behavior of ancient organisms). At still other levels, behavior is random (and therefore uninteresting). To interpret a macromolecular behavior, one must understand how the behavior arose. This understanding comes not from the study of a single behavior in a single protein, but from comparative studies of the behaviors of many proteins. The explosion in structural data makes such comparisons easier today than every before.

It is convenient to discuss enzymatic behaviors in terms of variable and conserved traits. Further, one speaks of enzymes related by a common ancestor as "homologous," with conservation or variation occurring in the divergent evolution of the ancestor's descendants. Variation between homologous proteins is the norm; no gene is free from the effects of random mutation. This means that if variability is *not* seen, an explanation is in order, not the other way around.

When variation is observed in the behavior of homologous proteins, it can be the result either of adaptive variation or neutral variation (or neutral "drift"). The first refers to behavioral variation that influences the survival of the organism that contains the enzyme (the "host" organism); the second variability has no impact on survival. The first can be interpreted in terms of biological function; the second cannot.

If a trait is conserved among homologous proteins, its conservation must be explained. One possible explanation is that the proteins diverged only very recently. Close homology in time implies close similarity in sequence, which implies (via structural theory) close similarity in behavior. In other words, if insufficient time has passed for two sequences to have diverged, behavior also should not have diverged.

Alternatively, however, the trait may have been conserved because it is important for the survival of the host organism. In particular, if two proteins are sufficiently different in structure that a biochemical trait *could* be different, it *should* be different, unless the particular trait serves a selected function. In this case, proteins with different traits confer different "survival values" upon the host organism; those where the trait is different are less successful in their effort to survive and reproduce, and natural selection "constrains" the drift of the trait under these circumstances.

A third explanation is also possible. It is conceivable that alternative traits are equally satisfactory solutions to a particular biological problem, but that no mechanism exists for a protein with one trait to evolve to become one with the other without creating intermediate proteins that are selectively disadvantageous. The exclusive use of L-amino acids in translated proteins may be an example of this; presumably D-amino acids would serve as well, but there is no obvious path for neutral drift from an organism using one to an organism using the other.

For traits that are not conserved, two contradicting explanations are possible. On one hand, the variation may have no impact on survival; the variation then reflects neutral drift. On the other hand, the variation might reflect adaptation of two proteins to two different environments.

Our view of the function of a trait depends critically on its evolutionary status; only selected traits can be functional. However, deciding which evolutionary interpretation is correct often requires much research, model building, and argument [12–16]. As much as one might like to avoid such complications, the issue cannot be simply ignored. Without an understanding of evolutionary processes, bioorganic data cannot be interpreted. Unless they are interpreted, we cannot use them to guide the design of models that mimic proteins, the engineering of proteins themselves, or development of a functional understanding of living systems.

Historical and Functional Pictures
A single sequence by itself often appears to be a non-informative jumble of letters. Similarly, the physical and catalytic properties of an enzyme appear to be a collection of trivia. However, from a comparison of the structures and behaviors of many biological macromolecules, we can construct hypotheses that distinguish traits in proteins that are selected, conserved, or drifting. These distinctions assign an evolutionary status to each behavior of each protein, separate important information from unimportant information, and suggest hypotheses regarding the chemistry of the macromolecules that can be tested experimentally. The process is described in detail elsewhere [12–16]. Here we give an overview of the best established hypotheses, together with some data that support them.

2 Selection, Drift, and Conservation

2.1 Structure and Behavior

Comparison of the behaviors of enzymes from many sources suggests the following conclusions about the relationship between structure and behavior in enzymes.

Point Mutations Can Create Substantial Variation in Behavior
The properties of a protein in vitro can vary widely with only small changes in structure. Single amino acid substitutions in ribosomal proteins raise and lower the fidelity of translation [17]. Six mutations in alcohol dehydrogenase from horse liver significantly alter the substrate specificity of the enzyme [18]. A single amino acid substitution in 5-enolpyruvylshikimate-3-phosphate synthase alters the ability of the enzyme to be inhibited by the herbicide glyphosphate [19]. A single amino acid change alters antigenicity and replication in an influenza virus [20]. Point mutations introduced by recombinant DNA technology have been found that completely destroy catalytic activity, increase or decrease stability of a protein, and alter regulatory properties, substrate specificity, or biological function [8]. An impressive collection of mutants of the

bacterial protease subtilisin have illustrated this for a protein with commercial importance [21].

However, Most Point Mutations Have Little Impact on Behavior

For example, Miller and his coworkers collected nearly one hundred nonsense mutations (those introducing a stop signal into the gene) in the gene coding for the lac repressor protein [22]. The majority of these had no detectable impact on behavior. More recently, Fersht and his colleagues have introduced many dozen mutations into the active site of tyrosine aminoacyl-RNA synthetase. Even though most of these mutations were at the active site, the impact on behavior of many was 'small' [8].

Even in Cases Where a Point Mutation Causes a Substantial Change in Behavior, the Behavioral Change Generally Can Be "Suppressed" By Alteration at Another Site

Certain random mutations in staphylococcal nuclease prepared by Shortle and his coworkers lowered the stability of the folded form of the protein [23, 24]. The effects of these mutations on enzymic stability could be suppressed by mutations at other positions in the sequence. Several of these "second site suppressors" were "global" i.e., they could improve the stability of the protein damaged by mutations at many other sites, Interestingly, one of these global suppressors is present in a naturally occurring nuclease [25].

Analogously, replacement of the catalytically important glutamate by aspartate in triose phosphate isomerase produces a mutant enzyme 10,000 fold poorer as a catalyst. Mutation at a second site restored a large fraction (100 ×) of the catalytic activity [26, 27].

The Relation Between Behaviors and Structure Can Be Scaled

The divergent behaviors of homologous proteins from nature suggests that alteration of some behaviors requires fewer structural alterations than others. For example, altering kinetic behavior appears to require fewer mutations than altering quaternary structure, substrate specificity, stereospecificity, catalytic mechanism, and gross tertiary structure, in that order (Table 1) [12].

Two examples illustrate these scales. Pancreatic and seminal RNAses have identical amino acids in 81% of the positions [28]. Yet their quaternary structures are different (one is a dimer, the other a monomer) [29], their substrate specificities are different (one acts on single stranded nucleic acid, the

Table 1. Specific examples relating behavioral and structural divergence

Variable by Point Mutation
Kinetic properties: kcat, KM, internal equilibrium constants
Regulatory properties, allosteric inhibition and activation
Thermal stability, substrate specificity, solubility, biological activity
Stereospecificity
Cofactor stereospecificity in yeast ethanol dehydrogenase

Table 1. (continued)

Variable with 10% sequence divergence
Substrate specificity
- Ethanol vs sterols (liver alcohol dehydrogenase) (2%)

Variable with 20% sequence divergence
Quaternary structure
- Seminal (dimer) and pacreatic (monomer) ribonuclease (19%)

Variable with 30% sequence divergence
Number of introns in gene
- Lysozyme

Variable with 40% sequence divergence
Substrate specificity[a]
- 2-Oxoglutarate dehydrogenase vs pyruvate dehydrogenase 45%

Variable with 50% sequence divergence
Substrate specificity[b, c, d]
- cycloisomerase 1 vs clc B muconate cycloisomerase
- tryptophan hydroxylase vs phenylalanine hydroxylase (49%)
- tyrosine hydroxylase vs tryptophan hydroxylase (55%)
- phenylalanine hydroxylase vs tyrosine hydroxylase (55%)
- acetylcholinesterase vs butyrylcholinesterase (45%)

Mechanistic differences[e]
- Superoxide dismutases (Mn vs Fe) 49%

Reaction type[f]
- Phosphoglycerate mutase vs diphosphoglycerate mutase (49%)

Variable with 60% sequence divergence
Intradomain disulfide bonds
- Mammalian RNase vs turtle RNase and angiogenin (60%)

Reaction type[g]
- Tryptophan synthetase vs threonine synthetase (58%)

Variable with 80% sequence divergence
Reaction types[h]
- Eukaryotic repressor qa-1S (*Neurospora crassa*) vs Shikimate dehydrogenase (*S. cerevisiae*) 75%
- Fumarase vs aspartase (76%)

Variable with >80% sequence divergence
Mechanistic differences and stereospecificity[i]
- Alcohol dehydrogenases (Zn^{++} vs no Zn^{++})

"Essential" active site residues diverge
- Lysozymes (homology by X-ray structure)

Reaction type[j, k]
- Malate synthase (cucumber)/Uricase (soybean nodule)/glycolate oxidase (spinach) (marginal alignment)
- Fructose-2,6-bisphosphatase vs phosphoglycerate mutase (active site peptides)

Different binding types
- Corticosteroid binding globin vs serine protease inhibitors

[a] Bradford AP, Aitken A, Beg F, Cook KG, Yeaman SJ (1987) FEBS Letters 222: 211;
[b] Frantz B, Chakrabarty AM, (1987) Proc Natl Acad Sci 84: 4460;
[c] Grenett HE, Ledley FD, Reed LL, Woo SLC, (1987) Proc Natl Acad Sci 84: 5530;
[d] Sikorav J-L, Krejci E, Massoulie J (1987) EMBO J 6: 1865;
[e] Schinina ME, Maffey L, Barra D, Bossa F, Puget K, Michelson AM (1987) FEBS Letters 221: 87;
[f] Shanske S, Sakoda S, Hermodson MA, DiMauro S, Schon EA (1987) J Biol Chem 262: 14612;
[g] Parson CA (1987) Proc Natl Acad Sci 84: 5207;
[h] Anton IA, Duncan K, Coggins JR (1987) J Mol Biol 197: 367;
[i] Jornvall H, Hoog J-O, Bahr-Lindstrom von H, Vallee BL (1987) Proc Natl Acad Sci 84: 2580;
[j] Volokita M, Somerville CR (1987) J Biol Chem 262: 15825;
[k] Pilkis SJ, Lively MO, El-Maghrabi MR (1987) J Biol Chem 262: 12672

other prefers double stranded nucleic acid) [30], and their biological activities are different (seminal RNase has potent antitumor activity, pancreatic RNase has no antitumor activity) [31,32]. However, other behaviors are much the same. "Angiogenins," proteins thought to have a role in the vascularization of solid tumors, have sequences that are only 40% identical to bovine pancreatic RNase [33,34]. At this level of sequence divergence, the number of disulfide bonds has diverged; angiogenins have three, mammalian pancreatic RNAses have four.

Alcohol dehydrogenases show still greater divergence in behavior. The three isozymes from yeast (95% identical) of alcohol dehydrogenases have different substrate specificities and stabilities [35]. The enzymes from yeast and horse liver (40% identical) have grossly different substrate specificities, kinetic properties, and quaternary structures [36]. Glucose dehydrogenases and ribitol dehydrogenases (25% identical) [37] catalyze somewhat different types of redox reactions. Glucose dehydrogenase catalyzes the oxidation of a hemiacetal, while ribitol dehydrogenase catalyzes the oxidation of a simple alcohol. The alcohol dehydrogenases from *Drosophila* and yeast (with 25–30% sequence similarities in one domain only) catalyze the same reaction, but with different mechanisms (metal versus no metal), substrate stereospecificities, and cofactor stereo-specificities [38].

The numbers in Table 1 are an upper limit to the amount of sequence divergence that is needed to produce a specific divergence in behavior. For example, trypsinogen and chymotrypsinogen have sequences that are 39% identical, yet have different substrate specificities. Thus, fewer than 61% of the residues must be changed to change substrate specificity. How many of the changes are necessary for the difference in substrate specificity is not known from comparisons of natural enzymes, and must be determined by site-specific mutagenesis [39]. What is remarkable, however, is the apparent fact that virtually any enzymatic behavior can be changed by an accumulation of a sufficient number of point mutations.

2.2 Structure and Selection

Comparisons of natural enzymes suggests the following conclusions about the relationship between structure and selection in proteins.

Point Mutations That Only Slightly Influence the Structure of Proteins Can Nevertheless Be Selectable

In the wild, the gene for the alcohol dehydrogenase from *Drosophila* can be found in many different structural forms [40]. Given the level of structural polymorphism in the gene, one would expect several dozen variants of the protein in the wild. In fact, only two are seen, and these variants correlate with latitude and altitude. The remainder of the structural variation at the genetic level is "silent"; it occurs in the third base of codons or in non-functional parts of introns. This strongly argues that virtually any structural variation in this protein is sufficiently disadvantageous that it is removed by natural selection whenever they occur.

However, Such Stringent Selection Does Not Apply to All Macromolecular Structures

Certain genetic structures are not functionally constrained from drifting. For example, pseudogenes are DNA sequences that resemble sequences for catalytically active proteins but which cannot themselves code for an active protein [41]. Pseudogenes are often truncated versions of active genes or contain stop codons internal to the coding sequence, and therefore cannot produce active proteins. They are believed to arise from a duplication of an active gene [42]. The structure of pseudogenes drifts rapidly (10^{-9} to 10^{-8} changes/site/year) [43–45]. Similarly, codon usage in higher organisms drifts rapidly, suggesting that natural selection does not strongly influence codon choice.

Drift in some coding sequences is only weakly constrained by natural selection. For example, the rate of divergence of structure in albumins is on the order of $5–7 \times 10^{-9}$ changes/site/year, comparable to the rate of drift of pseudogenes. This suggests that wide variations in the structure of albumin have little impact on the fitness of the host organism. Consistent with this view is the fact that albumins apparently perform no selected catalytic function, and many genetically transmitted variations in the structure of albumin in humans are not associated with clinical symptoms [46].

Likewise, the structure of the C peptide of pre-proinsulin (the portion that is removed proteolytically to create active insulin) diverges at a rate of 7×10^{-9}/site/year [47]. The sequence of fibrinopeptides diverges 6×10^{-9}/site/year [48]. That of alpha-fetoprotein gene diverges 1.5×10^{-9}/site/year [49]. In each case, the peptide is discarded before the protein becomes functional, and the rapid rate of drift is consistent with few functional constraints on the structure of the peptide.

Finally, the rate of divergence of the eye lens protein in the blind mole rat is faster than in sighted animals, suggesting that some of the functional constraints on drift are missing in the blind animal [50]. However, the rate of drift in the blind animal is still slower than that of pseudogenes, suggesting that some functional constraints on drift remain.

Natural variation in the structure of a single enzyme within a population (polymorphism) also suggests (but does not prove) that at least some structural variation is neutral [51]. For example, an esterase in *Drosophila* displays considerable structural polymorphism at the level of the protein. In man, many naturally occurring variants of hemoglobin are not associated with disease [52, 53].

Positive Adaptations Are Now Known

The structure of a few proteins diverges faster than pseudogenes, suggesting that structural variation optimizes proteins adapting to new or rapidly changing environments. For example, growth factors [54] and protease inhibitors [55] are rapidly diverging, suggesting that the structural variation reflects positive adaptation for new function in recent evolutionary history.

Convergent evolution of sequence (as opposed to well-known examples of convergent evolution of behavior and tertiary structure) may occasionally occur [56]. The sequences of a lysozyme from a recently evolved branch of ruminant primates is more similar to the sequences of lysozymes from ruminant artiodactyls (e.g. cow) than from non-ruminant primates. As lysozyme in ruminants serves a special role in digesting the walls of bacteria growing in the rumen, the discovery of sequence convergence in lysozymes from divergent lines of mammals is strongly suggestive of functional adaptation of the sequence itself for a particular biological function.

Scales of Selectability Can Be Constructed

The degree to which a particular biological macromolecule is under selective pressure depends on its function, its position in the hierarchy of information flow, and the organism. Here again, scales can be constructed.

1) The impact of a structural change on survival value often follows simple chemical intuition. Amino acid substitutions that conserve charge, size, or polarity appear more likely to be neutral than non-conservative substitutions. So are alterations on the surface of a protein. Thus, "conservative" substitutions are observed more frequently that "non-conservative" substitutions, even after correcting for biases in the genetic code [57]. Likewise, surface residues diverge faster than residues whose side chains lie inside a folded protein.
2) A second scale relates impact on survival to the position of the structural perturbation in the hierarchy of information transfer. Unused DNA in a genome is less "expensive" than expression of unused protein [58], as the cost of replication of a silent gene (in ATP units, for example) is much lower than for synthesizing many protein molecules from one gene. Most expensive is the undesired catalytic activity of an expressed protein. An enzyme catalyzing an irreversible reaction can dissipate enormous quantities of free energy, and undoing the damage can be quite expensive. The catalytic activity of an undesirably active protein can cost 10^{12} times more than simply carrying the DNA for the protein silently in the chromosome [16]. Qualitatively, such a scale is consistent with what is known about genetic regulation and evolution in microorganisms, although quantitative measurement of the "cost" associated with the synthesis of unused proteins remains problematic [59–62].
3) The impact of a mutation on survival depends on the size and complexity of the host organism. Viruses appear to be more sensitive to genetic variation in a single protein than prokaryotes, single cell eukaryotes, and multicellular eukaryotes, in that order. This is not surprising, as variation in behavior of a specific magnitude represents a larger fraction of the total metabolism of a bacterium than of a mammal. Further, much of the survival of a multicellular animal presumably depends on macroscopic physiology, a factor that influences the survival of bacteria far less.

4) The extent to which structural variation escapes the attention of natural selection depends on the chemical process in which the protein is involved. For example, because the geometric requirements for binding are chemically less stringent than for catalysis, binding proteins should display greater non-selected structural variation than catalytic proteins.

Further, the "difficulty" of catalyzing particular reactions can be estimated either by using organic chemists' intuition or less subjective measures such as the stereoelectronic demands on a reaction [63]. For example, the antibonding orbital accepting electron density from an attacking nucleophile is larger in an esterase than in an amidase. Based on this observation, structural variation in esterases is expected to influence rate of reaction (and hence is more likely to be neutral) than structural variation in amidases. Likewise, redox reactions involving electron transfer (e.g. perhaps xanthine oxidase) are geometrically less demanding than reactions involving hydride transfer (e.g. alcohol dehydrogenase). The more abundant (and apparently non-selected) polymorphism in xanthine oxidase from *Drosophila* than in corresponding alcohol dehydrogenases is consistent with this notion [51].

Scales permit the construction of a fortiori arguments about the selectability of behaviors for which we may have no specific data. Thus, if point mutations in alcohol dehydrogenase from *Drosophila* are the result of natural selection, one can argue a fortiori that multiple forms of alcohol dehydrogenase in yeast are also.

These scales can be applied to the problem of codon usage. The genetic code is degenerate; codon selection therefore can have no impact on the behavior of a protein after the gene is translated. Nevertheless, variations in codon usage influence the relative survival of variants of Q-beta virus in populations [64]. This suggests that codon selection in viruses reflects selective pressures. In bacteriophage T7, bias in codon use also suggests that codons choice is a selected trait [65].

In bacteria and yeast, the choice of codon correlates with the abundance of tRNA [66–70]. Again, the variation in codon usage is presumed to be functional, perhaps influencing the level of gene expression. In higher organisms, the rate of divergence in codon use is only slightly slower than the rate of drift in the structure of pseudogenes or introns [42]. Thus, codon usage does not appear to be selected, although there are many suggestions to the contrary [71–74].

2.3 Selection and Behavior

Comparisons of natural enzymes suggests the following conclusions about the relationship between behavior and selection.

In Behaviors That Can Be Altered By Point Mutation,
Patterns of Behavior Consistent With Functional Models
Reflect Functional Adaptation

This fact is best illustrated by kinetic parameters, which (vide supra) can be greatly altered by single amino acid substitutions. If identical kinetic parameters are observed in enzymes with divergent sequences, these parameters must be directly adaptive. An upper limit can be set for the tolerance of natural selection to variation in these parameters. For example, the k_{cat}/K_M values for certain enzymes remains within a factor of 10 of the second order diffusion rate constant in physiological media in widely divergent enzymes [12, 14]. Michaelis constants generally are within an order of magnitude of physiological substrate concentrations; again, this observation holds over a range of widely divergent enzymatic structures. As k_{cat} and K_M *could* drift with small changes in primary structure were they not functionally constrained, natural selection must constrain the drift of kinetic behavior by at least a factor of 10. Of course, as variation within this range might itself be adaptive, selective constraints could be still more stringent.

Physical properties also appear to be finely tuned. For example, site directed mutagenesis studies suggest that triose phosphate isomerases with increased stability are accessible by point mutation [76]. Nevertheless, triose phosphate isomerases from organisms living in normal environments generally display only moderate thermal stability. This suggests that a certain degree of thermal instability is an adaptive trait in this enzyme.

In some cases, natural selection appears to control behavior very precisely. For example, the internal equilibrium constants (reflecting the free energy difference between enzyme-bound substrates and enzyme-bound products) of an enzyme is expected on theoretical grounds to be "downhill" in the direction of metabolic flux in an enzyme, as this arrangement provides the enzyme with the greatest catalytic efficiency under a particular set of physiological conditions [77]. This notion has been tested using the isozymes of lactate dehydrogenases, one from muscle (where the flux is in the direction of lactate), and one from heart (where the flux is in the direction of pyruvate). The internal equilibrium constants of the two lactate dehydrogenases are different in the direction predicted by theory, and appear to be "tuned" to within 0.4 kcal/mol [78].

Natural selection appears to have finely tuned the kinetic behavior of lactate dehydrogenases as well. Lactate dehydrogenase from the fish *Fundulus heteroclitus* is polymorphic, and the two forms have different temperature optima [79]. The relative abundance of the two forms correlates closely with the temperature of the water in which the fish is found. Likewise, the temperature optima of lactate dehydrogenases from cow and fish correlate with the temperature in the environment where the enzyme was adapted [80]. This suggests that the kinetic parameters of this enzyme are finely tuned to the natural environment of the fish, and similar arguments can now be found for other enzymes [81–86].

Recently, site directed mutagenesis has been used to directly demonstrate a correlation between structure and selectability in yeast alcohol dehydrogenase. Mutants of yeast alcohol dehydrogenase that barely alter the properties (either kinetic or physical) of the protein in vitro, when returned to yeast lacking wild-type alcohol dehydrogenase genes, have been found to alter the rate of growth of the yeast up to 50% [78, 87]. Although the environment in which the yeast was grown was artificial, the mutant genes were not chromosomal, and the rate of growth is highly sensitive to precise conditions, small perturbations in the behavior of a single protein appear to be able to have substantial impact on the rate of growth of an organism.

"Semi-random" mutagenesis (mutagenesis not deliberately designed to examine evolutionary issues) also provides information concerning which behaviors are selected and which are not. Random mutagenesis searches "structure space" in the region immediately around the structure of the natural protein. Thus, it provides a view of the properties of those enzymes that natural selection clearly had access to, but did not "select", and the position of the behavior of the natural protein in the distribution of behaviors of the mutants can be informative about the process of natural selection. As more mutations are introduced, the behavior of the protein becomes more like the behavior of a protein with random sequence.

These points are illustrated in Fig. 1. The first graph shows the distribution of the k_{cat} values of a random set of proteins. The distribution is hypothetical, but not unreasonable. Most of the random peptides have very little catalytic activity, and only a few have a high catalytic activity. The second graph shows the distribution of k_{cat} values for a set of single point mutants of the native enzyme expected if k_{cat} is a maximized trait. As more point mutations are introduced, the distribution becomes more like the distribution of the random sequence (graph 3).

If k_{cat} were maximized, natural selection would have produced a protein with the locally highest value; this implies that all point mutations will have lower values for k_{cat}. Conversely, should one observe that the k_{cat} values of a set of point mutants are all lower than that of the native protein, one can conclude that k_{cat} is a maximized trait.

This contrasts with the distribution of traits expected in point mutants should a trait not be maximized. Here, the distribution of behavior will be evenly distributed around the behavior of the native enzyme, as illustrated in the fourth graph.

Such models assume that different behaviors in an enzyme can vary independently. Experiments suggest that this assumption is not a bad approximation. For example, Fersht's laboratory has provided a wealth of detailed kinetic data on mutant aminoacyl-t-RNA synthetases [8]. These mutants were not designed to test evolutionary questions, so may be regarded as "semi-random" from our point of view. As expected for a maximized trait, k_{cat} values of the mutants are distributed towards lower values. K_M values for these mutants are more evenly distributed around the wild type, suggesting that this parameter, if optimized, is not maximized.

Small perturbations in structure can greatly alter many macromolecular behaviors, including kinetic properties, thermal stability, substrate specificity, and regulatory behavior. When these behaviors are conserved in widely divergent enzymes, conservation must be explained with functional models. Bioorganic data collected on these traits are best interpreted functionally, not

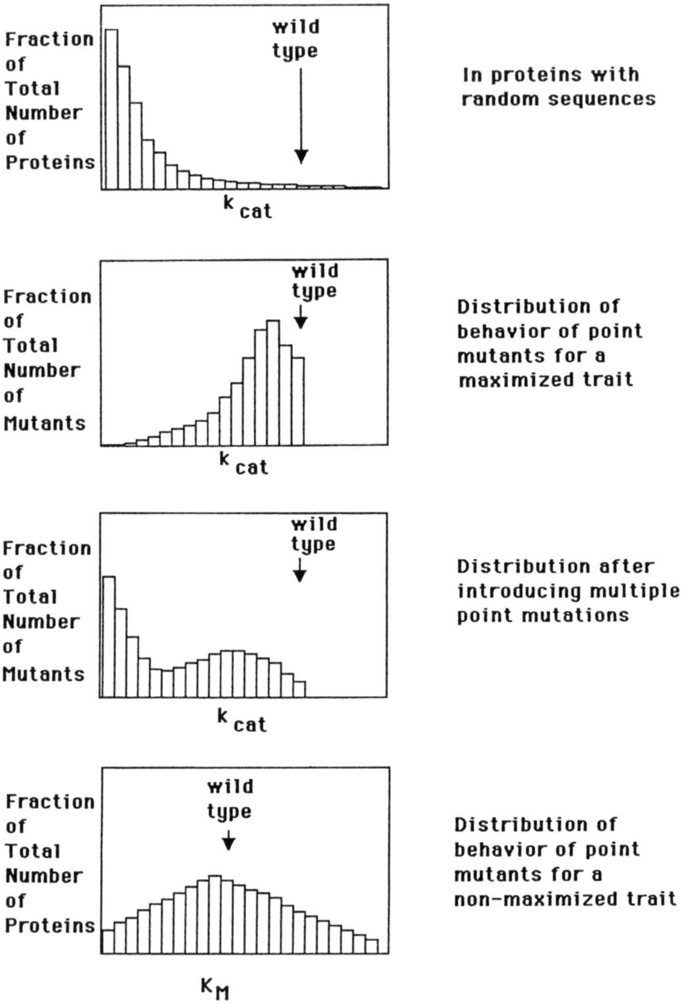

Fig. 1. The distribution of behaviors of randomly created mutants suggests whether or not an enzymatic behavior is maximized by natural selection. From reference 16

historically. Their exact values reflect natural selection, not drift, and not the previous history of the protein.

New Behaviors Can Arise in a Protein Simply By Changing a Few Amino Acids
The evolution of new function in a protein appears to be rapid under conditions where it can confer selective advantage. New catalytic activities can emerge by the introduction of a few point mutations into an existing enzyme as a selective response to an environmental challenge [88–94]. Three cases are worth mentioning in detail.

When *E. coli* was grown in the presence of maltose, eighteen mutants were found where the lactose transport protein had acquired the ability to transport maltose [95]. The mutations were not randomly distributed throughout the protein sequence, but rather occurred either at position 177 or 236. Ala-177 had become either Val or Thr, and Tyr-236 had become Phe, Asn, Ser, or His. Variants at position 177 retained their ability to transport lactose, while variants at 236 no longer effectively transported lactose. These results provide an interesting insight into the first steps in the adaptation of this protein. There appear to be only two "dimensions" in structure space [12] along which a lactose carrier can evolve to transport maltose by a single change. In one dimension, lactose and maltose transport are interdependent; in the other, they are independent.

In a second case [96,97], the gene for beta-galactosidase was deleted from *E. coli* and the organism grown with lactose as a carbon source. A new beta-galactosidase then evolved by introduction of 2 point mutations into another (as yet undefined) protein. The kinetic behavior of the intermediate proteins have been studied. The evolution was clearly adaptive and a linear relationship was found between growth rate (with lactose as a carbon source) and the k_{cat}/K_m for hydrolysis of lactose of the evolving gene [96,97].

Finally, a convincing case is now available for the rapid evolution (approx. 30 years) of a new chorismate mutase activity in a laboratory strain of *Bacillus subtilis* from which the original chorismate mutase had been deleted [98]. The new chorismate mutase arose from a binding site for prephenate, the product of the chorismate mutase reaction. The new enzyme is 50 fold slower than the deleted chorismate mutase, but still manages to effect a rate enhancement of 4×10^4.

The emergence of a catalytic site from a binding site in such a short time suggests that it is relatively easy to evolve a new enzyme, provided that a protein performing other functions already exists that contains part of the information needed for assembling the new enzyme. This is a laboratory illustration of a "deletion–replacement" event (vide infra). Such events presumably are common in macromolecular evolution, and will be important in our discussion below.

2.4 Conclusions

The relationships between structure and behavior summarized above present a picture of an easily adaptable protein. Starting with a protein of random sequence capable of forming a stable tertiary structure, a protein with virtually any behavior seems to be only a few point mutations away. Further, even if a route from one structure to another has an intermediate with an unacceptable behavior, suppressor mutations can create an alternative route that does not proceed via intermediates with unacceptable properties.

These general statements correspond to hypotheses in specific cases. Based on direct experimental evidence, many enzymatic behaviors can be assigned as

Table 2. Candidates for selectable macromolecular traits

Stereoselection between diastereomeric transition states
NADH-dependent redox reactions
Phosphoryl transfer reactions
Addition reactions to olefins
Internal Equilibrium Constants
Kinetic parameters $\pm 5\%$
Stability/instability
Substrate specificity against compounds present physiologically
Candidates for Neutral Macromolecular Traits
Stereoselection between enantiomeric transition states
Decarboxylation of beta-keto acids
Pyridoxal-dependent decarboxylation of amino acids
Non-equilibrium dynamic motion in proteins
Substrate specificity against compounds not present physiologically

"selectable" or "non-selectable" (Table 2). Each of these hypotheses can be experimentally tested.

3 The Use of Functional Models in Enzymology Protein Engineering and Prediction of Tertiary Structure

Conclusions such as those presented above are especially interesting because they suggest experiments and new approaches in solving problems important in bioorganic chemistry.

3.1 The Use of Evolutionary Models in Protein Engineering

Evolutionary generalizations such as those discussed above allow a statement as to what kinds of protein engineering experiments are likely to succeed, and what kinds are likely to fail. In general, efforts to introduce into a protein a non-selected trait are likely to succeed, while efforts to engineer selected traits into proteins are likely to fail. Conversely, it should be relatively easy to remove a trait that is desired by natural selection, and relatively difficult to remove a trait present in a protein that is not desired by natural selection.

The rationale for these rules is simple. If a selectively undesirable trait remains in a natural protein, this implies that a billion years of evolution have failed to remove it and that it is intrinsically associated with the selected traits of the protein. In this case, it is unlikely that the protein engineer will remove it in the laboratory. Likewise, if a desirable trait has not appeared in a protein in a billion years, the engineer is not likely to introduce it in the next five.

In contrast, if a trait desired by the engineer is not desirable in the eyes of natural selection, this is sufficient reason for its absence in a natural protein. As virtually any behavior in a protein seems accessible by only a few point mutations, the engineer has a chance of introducing the trait in this case. Conversely, a trait present in a natural protein that is desired by natural selection is placed and maintained there by constant selection for a few critical amino acids; changing those amino acids should remove the trait.

For example, thermal instability appears to be a trait desired by natural selection (vide supra). Thus, improving the thermal stability of a protein by point mutation should be possible, a speculation that is now documented in several cases [99, 117, 118].

This rule can be applied to evaluate the possibility of engineering ribulose diphosphate (RuDP) carboxylase [100]. This enzyme has an undesirably broad substrate specificity; it catalyzes the reaction of ribulose diphosphate with both carbon dioxide and oxygen (Fig. 2). The first reaction is desired; the second leads to the oxidative cleavage of the substrate. The oxidative cleavage wastes substrate, and simple calculations suggest that if plants had a RuDP carboxylase with narrower substrate specificity (eliminating the oxygenase reaction), they would fix carbon dioxide with nearly double the efficiency, solving (at least temporarily) the world's food problems [100–102].

Whether or not one undertakes this challenge in protein engineering depends critically on whether one views the oxygenation reaction as a trait desired by natural selection. If it is, constructing a ribulose diphosphate

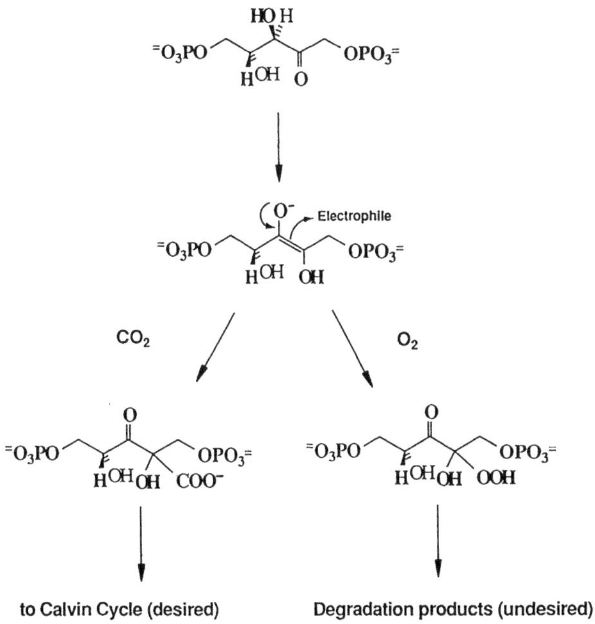

Fig. 2. Ribulose 1,5-diphosphate carboxylase forms an enolate intermediate, which reacts with either carbon dioxide (presumably the desired reaction) or oxygen (presumably an undesired side reaction intrinsic to the chemistry of the enolate intermediate)

carboxylase that lacks the trait should be relatively easy. If it is not, engineering the enzyme to remove the trait is likely to be extremely difficult. If it were possible to remove the oxygenation reaction via point mutation, evolution would have done so long ago if it were selectively undesirable.

Comparison of various RuDP carboxylases suggests that the ratio of oxygenation to carboxylation is conserved [103]. This suggests either that the trait is adaptive, or that it is intrinsically coupled to another adaptive trait (e.g. the carboxylation reaction). However, the ratio is also conserved in RuDP carboxylases from anaerobic bacteria [103]. The last point is surprising; if oxygenation were a selected trait in aerobic organisms, it should have been lost through drift in anaerobic organisms. This suggests that the oxygenase reaction is an unavoidable side reaction intrinsic to the chemistry of carbon dioxide fixation, a plausible suggestion considering the chemical reactivity of the enolate that is a putative intermediate in the reaction (Fig. 2) [104].

Further, many succulents expend ATP to pump carbon dioxide to the RuDP carboxylase, increasing its concentration relative to oxygen, and therefore increasing the carboxylation reaction relative to the oxygenation reaction. Were the oxygenation reaction avoidable by point mutation of the enzyme itself, it seems unlikely that an energy-consuming alternative would have evolved.

This evolutionary perspective provides a warning to the many laboratories who are attempting to remove the oxygenase reaction in RuDP carboxylase by point mutation. If the analysis above is correct, the oxygenation reaction is selectively disadvantageous, yet remains in natural proteins because it is intrinsically coupled to the advantageous carboxylation reaction. What evolution has failed to remove on a geological time scale will be difficult to remove on the Ph.D. time scale.

An evolutionary view of stereospecificity in dehydrogenases has also guided engineering experiments. Stereospecificity is highly conserved in many dehydrogenases. If stereospecificity is not directly functional (as is widely believed), there must be a secondary constraint on drift that makes it difficult or impossible to reverse stereospecificity without producing inactive proteins. The most forceful statements of this position are due to Oppenheimer, who has argued that flipping the orientation of the nicotinamide ring (a flip that presents the opposite face of the cofactor to substrate, thereby reversing stereospecificity at the cofactor), is blocked by steric interactions between the carboxamide group of the cofactor and the side chains of the cysteine residues that serve as ligands to the zinc in the active site (Fig. 3) [105]. As the zinc is essential for catalysis, it follows from this argument that stereospecificity cannot be easily reversed without destroying catalysis. This argument is used to explain the high conservation of stereospecificity while denying a direct functional role for stereospecificity itself.

If Oppenheimer's conjecture is correct, it should be impossible to engineer a catalytically active alcohol dehydrogenase that accepts 5-methylnicotinamide as a cofactor. The methyl group at position 5 occupies the same space that would be occupied by the carboxamide group in the flipped conformer of the natural

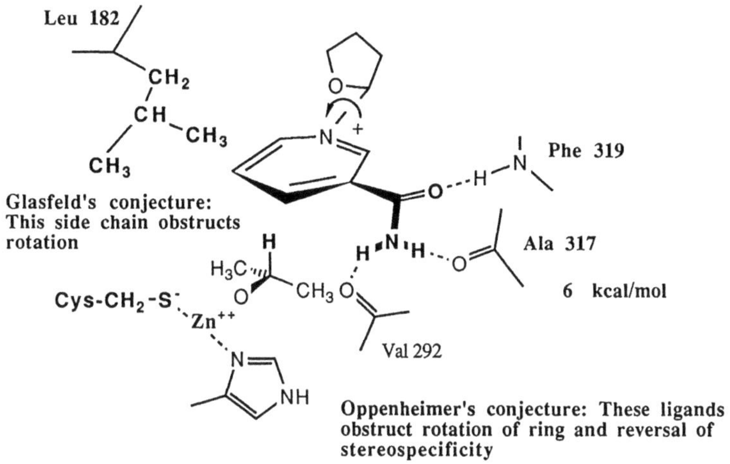

Fig. 3. Two possible structural features of the active site of alcohol dehydrogenase that might determine the stereospecificity of the enzyme with respect to cofactor

cofactor (Fig. 3). Thus, if Oppenheimer's conjecture is correct, the methyl group should also interact unfavorably with the ligands to zinc and disrupt catalytic activity (Fig. 4).

However, considerable evidence suggests that stereospecificity in many dehydrogenases is a directly selected trait [13, 106–109]. This implies that no special structural constraint is needed to explain the conservation of stereospecificity, in particular, one involving the ligands to zinc. Further, inspection of the crystal structure of alcohol dehydrogenase with a perspective enlightened by this view suggests that Oppenheimer's conjecture is incorrect; the steric obstruction for the ring flip comes not from the side chains of cysteines coordinated to zinc, but rather from residue 203 (in the horse enzyme), corresponding to residue 182 (a leucine) in the homologous alcohol dehydrogenase from yeast. This side chain is not obviously important for catalysis. In this view, a dehydrogenase with a smaller side chain at this position should retain catalytic activity, have decreased stereospecificity, and have the capacity to use 5-methylnicotinamide adenine dinucleotide as a cofactor.

These latter predictions were recently confirmed by Arthur Glasfeld and Elmar Weinhold in this laboratory. Leu-182 was replaced by Ala in yeast alcohol dehydrogenase [87]. The mutant is quite active catalytically, but its stereospecificity with respect to cofactor is four orders of magnitude lower than that of the natural enzyme. Further, the mutant accepts 5-methylnicotinamide as a cofactor; the natural enzyme does not (Fig. 4).

This successful engineering experiment would not have been attempted by one who had misread the evolutionary status of this particular macromolecular trait. This underscores the importance of evolutionary hypotheses for a bio-

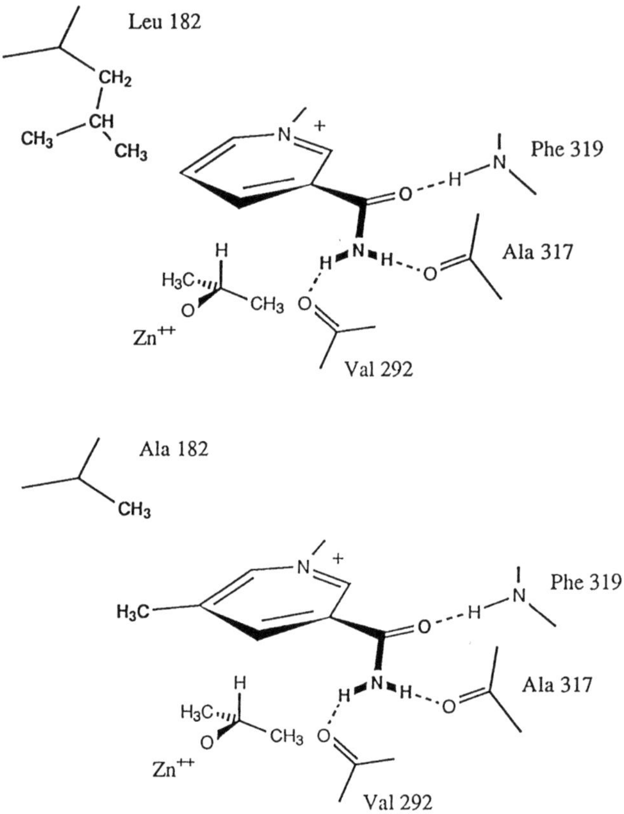

Fig. 4. Replacement of residue 182 in Yeast alcohol dehydrogenase (corresponding to residue 203 in the mammalian enzyme) lowers the stereospecificity of the enzyme and allows it to accept 5-methylnicotinamide adenine dinucleotide as a cofactor

organic chemist, both to understand the proteins he studies, and to manipulate them using site-directed mutagenesis techniques.

3.2 Evolutionary Perspectives and Protein Structure

Evolutionary considerations can also assist in interpreting the structural details of proteins, and assist predictions of the structure of proteins given a set of homologous proteins. This analysis begins with the important hypothesis that a degree of instability is a selected trait in natural enzymes.

Certainly, the turnover of enzymes is important to the health of cells; mutants of *E. coli* that lack degradative enzymes grow poorly [110, 111]. Strictly, this implies only that protein instability in vivo is desired, not the thermal instability in vitro that normally concerns bioorganic chemists. However, thermal stability in vitro correlates roughly with physiological stability

[112]. Therefore, a degree of thermal instability in vitro is expected in well-adapted enzymes, especially enzymes that are found intracellularly under physiological conditions.

Mutagenesis studies also support this conclusion. Protein stability has proven to be one of the easiest properties to improve by site-directed mutagenesis (vide supra), especially when compared with efforts to improve k_{cat} or other evolutionarily maximized traits. While one might interpret these relative frequencies of success as evidence that bioorganic chemists understand better what structural features are responsible for thermal stability than they understand the structural features responsible for catalysis in a protein, more likely they reflect the fact that natural selection does not seek out maximum thermal stability (even a local one) in the evolution of a protein. Thus, the structure space [12] around the wild type protein contains many proteins that are more stable, and these are easily found, even in a random walk.

Insight into the chemical basis of thermal stability has come from studies on enzymes from thermophilic bacteria [113]. For example, thermophilic enzymes have low levels of asparagine; asparagine is easily deamidated, and deamidation of asparagine is an important first step in the irreversible denaturation of many proteins [114]. However, a statement that asparagines are removed from mesophilic enzymes to stabilize enzymes in thermophiles can be reversed. We can say as easily that asparagines are *introduced* into enzymes from thermophiles as a way of *destabilizing* an enzyme from a mesophile. A priori, one cannot say from these data whether the destabilizing interactions in the mesophilic enzyme arose by drift, or by adaptive selection for a less stable protein. However, from the arguments above, the second choice proves to be more valuable as a working hypothesis.

To illustrate, we compare the sequences of two archaebacterial phosphoribosylaminoimidazole carboxylases, one adapted for growth at high temperatures (*Methanobacterium thermoautotrophicum*), the other adapted to growth at low temperatures (*Methanobrevibacter smithii*) [115]. The enzymes catalyze identical reactions in identical pathways in two organisms that are quite similar in overall metabolism. Nevertheless, the sequences of the two proteins are 55% different. Some of the sequence divergence in the two proteins is undoubtedly due to drift. However, some must be functionally adaptive, and many of the adaptive differences must relate to the different temperatures of the environments of the two proteins.

Even without a crystal structure, information can be extracted as to what those adaptive differences might be. Table 3 shows a matrix of interconversions between the mesophilic enzyme and the thermophilic enzyme. Strikingly, in the mesophilic enzyme, many of the arginines of the thermophilic enzyme have been replaced by lysine. A similar pattern was noted by Zuber in lactate dehydrogenase from mesophilic and thermophilic organisms [116].

Lysine has a lower pK_a than arginine. Further, the positive charge in arginine is delocalized over 2 atoms in a geometry that mirrors the delocalization of the negative charge in carboxylate groups of aspartate and glutamate.

Table 3.

									Mesophilic												
		R	K	E	D	Q	N	H	S	C	T	P	G	A	V	I	L	M	F	W	Y
	Arg	5	11	2										2		1	1				
	Lys	1	6	1			3				2				1	1	1				
t	Glu	2	6	3	4	2	2					1					1		1		
h	Asp			3	7		4		1				1				1				
e	Gln	1		1		1													1		
r	Asn					2	5										1				1
m	His					1		3				1									
o	Ser	1	2	2			2		5	1	3	2		2	3						1
p	Cys			1					1	1											
h	Thr								2		4				1						
i	Pro		1	2			2		1		1	15	1		1						
l	Gly	1	1	2			2		3				18	1							1
i	Ala			1					5		1		8	14			1				
c	Val			1							1			1	23	7	2				
	Ile										1				2	19	7				
	Leu				1		2		1						3	2	14	1			
	Met			2							1			1	1	2	2	6			
	Phe										1					2	1		3		3
	Trp						1														
	Tyr						1	3	1						1						2

For both reaons, lysine should form weaker salt bridges with carboxylate anions than arginine. In this view, the substitution of arginine by lysine in the mesophilic enzyme is an adaptive effort to lessen the conformational stability of the mesophilic enzyme.

A second interesting change is that 8 alanines in the thermostable protein are glycines in the mesophilic protein; only 1 glycine becomes an alanine. Statistically insignificant, but noteworthy nevertheless, is the conversion of a proline in the thermostable enzyme to a glycine in the mesophilic enzyme. Both changes substantially increase the entropy of the unfolded form in the mesophilic protein, thus destabilizing the folded form relative to it. Similar changes have been used (in the reverse direction) to increase the stability of a mesophilic protein by "protein engineering" [117, 118].

Asparagines appear in the mesophilic protein to replace other residues in the thermophilic enzyme. Here again, these replacements can be viewed as adaptive efforts to introduce instability into the mesophilic protein, as asparagine is easily deamidated.

3.3 Tertiary Structure Prediction

An evolutionarily comprehensive view of protein structure can provide other benefits to the biological chemist. One of the most important of these is an approach being developed for building low resolution structures of enzymes starting with a set of homologous proteins [119]. Methods for predicting the tertiary structure of proteins from sequence data alone are among the most desired in bioorganic chemistry, having been the targets of research for over two decades [120]. Although a measure of success has been obtained, predictions of secondary structure by "classical" methods remain less than 70% accurate, even when classical methods are averaged over a series of homologous proteins (presumed to have largely similar folded structures [4]) to filter out the noise [120]. This level of accuracy is generally insufficient to sustain the building of tertiary structural models but see reference 279.

As evident from the discussion above, substantial information relevant to tertiary structure of a protein is contained in the pattern of divergence and conservation of the sequences of homologous proteins. For example, lysines that are replaced by arginines in homologs from thermophilic organisms are likely to be involved in salt bridges. Glycines that become alanines in more stable proteins are likely to be found in ordered secondary structures such as helices. Glycines that are replaced by prolines in some proteins in more stable proteins are likely to be involved in turns. Asparagines that are replaced by other polar residues are likely to be on the surface of a protein, perhaps at quaternary contacts [78].

Each of these patterns is a clue that could be used to build a model for the tertiary structure of a set of homologous proteins. However, before a comparison of a set of homologous sequences can be used to predict tertiary structure, issues that are fundamentally evolutionary in nature must be resolved.

Inspection of a set of homologous proteins can detect only two things, conservation or variation. However, as discussed above, variation can be either neutral or adaptive, and these two types of variation have exactly opposite implications with respect to folded structure. Neutral variation is variation that influences no selectable behavior of the enzyme. Between closely homologous proteins, it therefore will occur primarily on the surface, as this is where variation has the smallest impact on behavior (note, of course, that this statement does not imply that all variation on the surface of a protein is neutral). Thus, naively, after inspecting an alignment of homologous proteins, one might assign positions displaying variability to the surface, and use these assignments as the beginning of a tertiary structural model for a protein (Fig. 5).

Such an algorithm for detecting surface residues is quite bad, and for obvious reasons. Adaptive variation is superimposed on neutral variation in a set of proteins and, in contrast to neutral variation, adaptive variation is *intended* to alter the behavior of a protein. It can therefore occur anywhere in the tertiary structure, even at the active site.

Further, as mentioned above, instability appears to be a selected trait in proteins. As there is an overabundance of intramolecular interactions available to a normal sized protein to stabilize a folded conformation [119], the average protein must violate a large number of folding "rules" to engineer a desired level of instability into its structure. This means that even if the biochemist were to learn perfectly the "rules" for folding proteins, he would still find it difficult to predict the tertiary structure of a protein from a single sequence amid all of the violations of the rules placed in the sequence by natural selection.

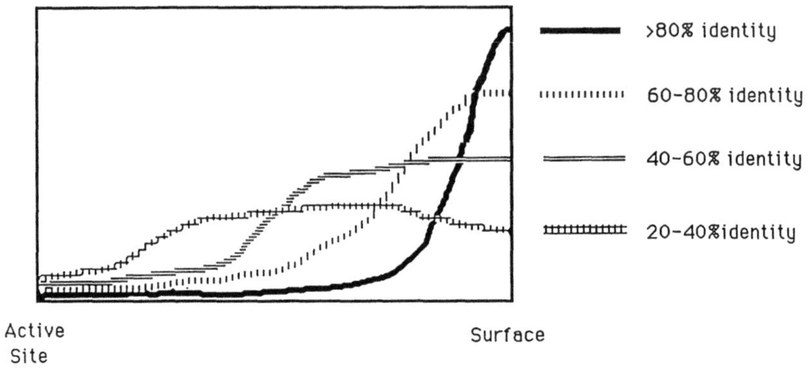

Fig. 5. A graph illustrating the distribution of neutral variation as a function of radial distance from the active site in proteins that have 80% sequence identity (heavy black line), and in proteins with greater sequence divergence

At first glance, this is discouraging. However, once these considerations were appreciated, successful algorithms can be developed for predicting secondary and tertiary structure in proteins from the pattern of conservation and divergence in a set of homologous sequences [119]. These now can be used for predicting the tertiary structure for many soluble enzymes in certain cases. Eight steps are involved in principle:

a) Construct an alignment of the sequences of a set of homologous proteins. The alignment should contain proteins whose sequences have diverged to the point where only 15–20% of the positions in the alignment are conserved across the entire alignment. However, the alignment should contain some pairs of proteins with similar sequences (approx. 80% identity).
b) Apply algorithms that extract tertiary structural information for individual residues from the alignment, in particular, those that assign surface residues, interior residues, and residues at the active site.
c) Divide the alignment into segments corresponding roughly to secondary structural elements using parsing algorithms.
d) Assign secondary structure to the segments using tertiary structural information extracted in step b.
e) Orient the secondary structural units around a focal point on the protein, usually (for enzymes) the active site.
f) Match the assembly of structural units, if possible, to a particular taxonomical class of proteins.

Limitations of space prevent a discussion of how alignments (point a), or the types of taxonomical classes in proteins (point f) are obtained [158]. However, some of the most useful algorithms for predicting the tertiary structure of a protein (points b, c, d, and e) are reviewed below, using a set of homologous alcohol dehydrogenases (Adh) as a vehicle for illustration.

To understand these algorithms, we must begin with some definitions. An alignment of homologous proteins serving analogous biochemical functions contains positions typically designated by two numbers, an alignment position number (which counts all deletions and insertions) and a crystal sequence number, in this case the sequence number of the residue in horse liver alcohol dehydrogenase (Adh), which is used here as a reference structure for illustration. A typical alignment is shown in Fig. 6 for 17 homologous dehydrogenases, and the alignment number (the first number in the Figure) will be used in the discussion below.

The alignment can be divided into subgroups having specific minimum pairwise identities (or MPI's). For example, when the proteins in a subgroup with an MPI of 85% are compared pairwise, the sequence identities between the

		Mammals								Plants				Fungi				
365	**351**	Pro	Pro	Pro	Pro	Pro	Pro	Pro	Pro	Pro	Pro	Pro	Pro	Gly	Gly	Gly	Pro	Pro
366	**352**	Phe	Phe	Phe	Phe	Phe	Phe	Phe	Phe	Phe	Phe	Phe	Phe	Leu	Leu	Leu	Leu	Phe
367	**353**	Glu	Glu	Glu	Glu	Glu	Glu	Glu	Asp	Ala	Ser	Ser	Ser	Ser	Ser	Ser	Gln	Ser
368	**354**	Lys	Lys	Lys	Lys	Lys	Lys	Lys	Lys	Glu	Glu	Glu	Glu	Thr	Ser	Glu	Asp	Thr
369	**355**	Ile	Ile	Ile	Ile	Ile	Ile	Ile	Ile	Ile	Ile	Ile	Ile	Leu	Leu	Leu	Leu	Leu
370	**356**	Asn	Asn	Asn	Asn	Asn	Asn	Asn	Ser	Asn	Asn	Asn	Asn	Pro	Pro	Pro	Pro	Pro
371	**357**	Glu	Glu	Glu	Glu	Glu	Glu	Glu	Glu	Lys	Thr	Lys	Lys	Glu	Glu	Lys	Gln	Asp
372	**358**	Gly	Gly	Gly	Gly	Gly	Ala	Ala	Ala	Ala	Ala	Ala	Ala	Ile	Ile	Val	Ile	Val
373	**359**	Phe	Phe	Phe	Phe	Phe	Phe	Phe	Phe	Phe	Phe	Phe	Phe	Tyr	Tyr	Tyr	Phe	Tyr
374	**360**	Asp	Asp	Asp	Asp	Asp	Asp	Asp	Asp	Asp	Asp	Asp	Asp	Glu	Glu	Asp	Glu	Arg
375	**361**	Leu	Leu	Leu	Leu	Leu	Leu	Leu	Leu	Leu	Leu	Tyr	Tyr	Lys	Lys	Leu	Leu	Leu
376	**362**	Leu	Leu	Leu	Leu	Leu	Leu	Leu	Met	Met	Met	Met	Met	Met	Met	Met	Met	Met
377	**363**	Arg	Arg	His	His	Arg	Arg	Arg	Asn	Ala	Leu	Leu	Leu	Glu	Glu	Glu	Gly	Asn
378	**364**	Ser	Ser	Ser	Ser	Ser	Ser	Ala	Gln	Lys	Lys	Lys	Lys	Lys	Lys	Lys	Gln	Glu
379	**365**	Gly	Gly	Gly	Gly	Gly	Gly	Gly	Gly	Gly	Gly	Gly	Gly	Gly	Gly	Gly	Gly	Asn
380	**366**	Glu	Lys	Lys	Lys	Lys	Lys	Lys	Lys	Glu	Glu	Glu	Glu	Gln	Gln	Lys	Lys	Lys
381	**366**	---	---	---	---	---	---	---	---	---	---	---	---	Ile	Ile	Ile	Ile	Ile
382	**367**	Ser	Ser	Ser	Ser	Ser	Ser	Ser	Ser	Gly	Gly	Ser	Ser	Val	Ala	Leu	Ala	Ala
383	**368**	Ile	Ile	Ile	Ile	Ile	Ile	Ile	Ile	Ile	Leu	Ile	Ile	Gly	Gly	Gly	Gly	Gly
384	**369**	Arg	Arg	Arg	Arg	Arg	Arg	Arg	Arg	Arg	Arg	Arg	Arg	Arg	Arg	Arg	Arg	Arg
385	**370**	Thr	Thr	Thr	Thr	Thr	Thr	Thr	Thr	Cys	Cys	Cys	Cys	Tyr	Tyr	Tyr	Tyr	Ile
386	**371**	Ile	Ile	Ile	Val	Val	Val	Val	Ile	Ile	Ile	Ile	Ile	Val	Val	Val	Val	Val
387	**372**	Leu	Leu	Leu	Leu	Leu	Leu	Leu	Leu	Ile	Met	Ile	Ile	Val	Val	Val	Leu	Leu
388	**373**	Thr	Thr	Met	Thr	Thr	Thr	Thr	Ile	Arg	Arg	Lys	Thr	Asp	Asp	Asp	Glu	Asp
389	**374**	Phe	Phe	Phe	Phe	Phe	Phe	Phe	Phe	Met	Met	Met	Met	Thr	Thr	Thr	Ile	Leu

Fig. 6. Part of the alignment of the 17 homologous alcohol dehydrogenases shown in the tree in Fig. 7. The dotted line divides alignment into subgroups with minimum pairwise identities (MPI) of 85%, the solid line subgroups with MPI = 50%.

least similar sequences is 85%; the identity between other sequences can, however, be much larger. The divergent evolution of the protein family can be represented on an evolutionary tree (Fig. 7); a subgroup represents a sub-branch of this tree. These subgroups are often given labels based on the organisms that are the sources of the enzymes in the subgroup. Thus, at an MPI = 85%, the subgroups within the alcohol dehydrogenase family are labeled "rodent", "maize", "human", etc. The dotted vertical lines in Fig. 6 divide the alignment into 9 subgroups with an MPI of 85%; the solid vertical line divides the alignment into 2 subgroups with an MPI = 50% (see also Fig. 7).

All subgroups with a particular MPI are collected together as a cluster of subgroups with a particular MPI. When the residue at position n in an alignment is the same in all members of a subgroup with an MPI = x, it is conserved across a subgroup with MPI = x". When the residue is conserved across an entire alignment, it is designated "Absolutely perfectly conserved" (or "APC"). An example of an APC residue is 384 in the alignment shown in Fig. 6. When the residue is conserved in all subgroups in a cluster with an MPI = x, but is different in different subgroups, the position is a split at an MPI = x. For example, position 369 is a "hydrophobic split at MPI = 50%." Splits can be found with respect to other properties of a subgroup. For example, a variability split is a position where the residue of some of the subgroups is conserved and in others variable. Position 374 is an example at MPI = 50%, with Asp conserved

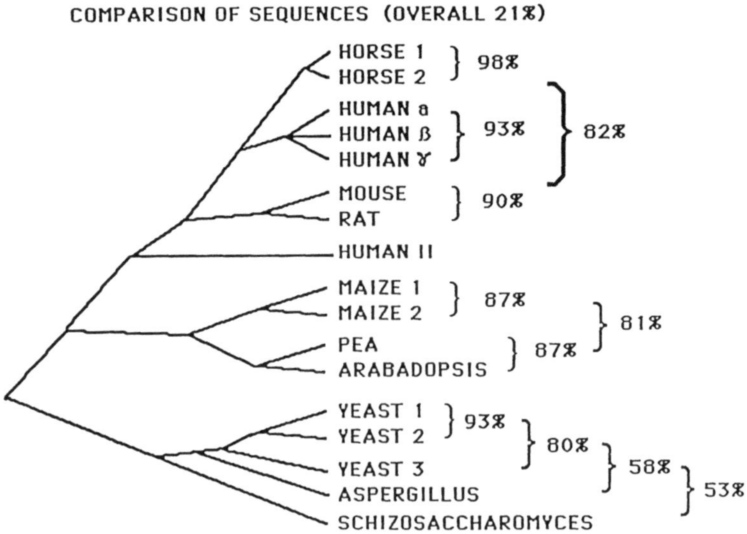

Fig. 7. An evolutionary tree showing the relationship between alcohol dehydrogenases whose sequences are used to predict tertiary structure. Five subsets of the alignment are used to predict surface residues: The horse and human enzymes (except Human Adh II), the rodent enzymes, the maize enzymes, pea and Arabadopsis Adh, and yeast Adh I and 2

in the branch of the tree including mammals and plants, but variability is observed (Glu, Asp, or Arg) in the fungal branch.

When a residue is not the same within a subgroup at MPI = x, the residue is variable at MPI = x. Variable residues may nevertheless share properties, including polarity (hydrophobic or hydrophilic, or compatible or incompatible polarities), size, structure disrupting properties, and so on. When comparing subgroups in a cluster, variability is said to be reflexive when some of the amino acids in one variable subgroup appear in other of the subgroups.

If a structural feature is found at adjacent positions in the alignment, it is said to be distributed over several alignment positions.

A set of simple algorithms allows one to predict whether the side chain of an amino acid residue at a particular position will lie on the surface of the protein or in the interior of the protein, depending on the pattern of conservation and the nature of the amino acids conserved. The most useful algorithm for assigning residues to the surface of a protein simply involves dividing the alignment into a cluster of subgroups with a particular MPI (85% is used first for this example), and assigning a residue to the surface if at least 2 of the subgroups in the cluster display internal variability and at least one protein has at that position either an Asp, Glu, Lys, Arg, or Asn.

For example, in Adh, there are 9 subgroups (Fig. 6, 4 of these contain a single protein) with an MPI of 85%. Inspecting the region of the alignment between alignment positions 365 and 389, only positions 377 and 388 fulfill both requirements, variability in at least 2 subgroups and having Asp, Glu, Lys, Arg, or Asn in this position in at least one protein. At position 365, the first (mammal) and fourth (plant) subgroups display variability, and Arg, Glu, and Asn all appear in at least one protein at this position. Likewise, at position 388, subgroup 1 and subgroup 5 display variability, and Arg, Asp, Lys, and Glu are found at least once in some protein at this position. Thus, these residues are assigned to the surface.

The rationale for this algorithm is simple. Variation at position 377 in a subgroup of mammalian proteins could be either neutral or adaptive, reflecting the need for two different protein sequences in two different mammalian environments. However, the variation in two different plant environments is not likely to be so parallel that variation at the same position would be adaptive in plants as well. Thus, the requirement that variability be observed in two subgroups before it is treated as a significant indicator of surface position will have the effect of "filtering out" adaptive variation that would be misassigned to the surface.

How well this filter works must be determined empirically. For Adh, this algorithm assigns 31 residues (corresponding to 8% of the total alignment) are assigned to the surface; 97% of the assignments are correct (Table 4).

By changing the MPI of the subgroups that are examined, the number of assignments can be increased, with a corresponding decrease in the reliability of the assignments. For example, looking for variability in more than two subgroups with an MPI of 80% will identify more surface residues, as there is more

Table 4. Assignment of surface residues by examining subgroups with different minimum pairwise identities

MPI (minimum)	Residues identified	% of alignment	% correct
90%	8	2	100
85%	31	8	97
80%	45	12	96
70%	70	20	96
60%	91	23	95
50%	95	24	94

variability within these classes generally. Indeed, this is the case (Table 4). With subgroups with an MPI of 80%, 45 residues are identified (12% of the total alignment) with 96% accuracy.

The optimal MPI is the one that gives the highest product (% of the surface residues identified) × (% accuracy). As is evident from Table 4, this optimum occurs for Adh with subgroups with an MPI of 50 or 60%. Using subgroups with higher MPI's lowers the number of assignments without greatly improving their reliability. Using subgroups with lower MPI's reduces the reliability of the assignments without greatly increasing their numbers.

Examining subgroups with an MPI of 50%, approximately 60% of the surface residues are assigned in Adh (Table 4). The reliability of the assignments does not change significantly with a smaller alignment, as long as the remaining proteins remain similarly distributed across the tree. However, the number of assignments does change. This is important, as it implies that by collecting more sequence data for a set of homologous proteins, one can increase the fraction of the surface successfully assigned. The significance of this point becomes more evident when one recognizes that one can easily clone genes of proteins 60% homologous to a protein with a known sequence, with the second gene serving as a probe for the first.

Positions that display variability in more than one subgroup but do not contain Asp, Glu, Lys, Arg, or Asn at in some protein are useful for other reasons. When these positions contain residues bearing functional groups (Cys, His, Gln, Ser, or Thr), they can only weakly be assigned to the surface. However, because the side chains of such residues often lie inside a structure, because they often require compensating hydrogen bonds from other residues on the inside, and because they are variable, they are often valuable for testing a model through covariant analysis (step g).

Further, positions that display variability in more than one subgroup but have hydrophobic residues conserved throughout cannot be assigned to the surface (and they rarely are on the surface with clusters of subgroups with MPI of 60% or less). However, they are good indicators of contacts between subunits and domains, depending on the MPI for which they are observed.

Analogous algorithms can make assignments for internal residues. Here, conservation (as opposed to variability) in subgroups with lower MPI (as opposed to higher MPI) is an indicator of an interior position. Further, hydrophobicity rather than hydrophilicity is an indicator of an interior position. A hydrophobic residues (Phe, Tyr, Trp, Met, Leu, Ile, Ala, Val) conserved in a subgroup with a MPI of 30% is a 100% reliable indicator of an interior position. Nearly as good is a "hydrophobic split" in a set of subgroups with an MPI = 50 or 60%. Position 369 (Fig. 6) is an example of a residue that is assigned as an interior residue based on such an algorithm (MPI = 50%).

Analogous algorithms can be constructed to assign residues to the active site. Residues bearing functional groups (Cys, His, Gln, Ser, Thr, Asp, Asn, Glu, Lys, or Arg) that are absolutely conserved across an entire alignment where the MPI is less than 30% are subject to extreme functional constraints on drift. Normally, such constraint can come only from the participation of the residue in catalysis in an enzyme. However, charged residues involved in salt bridges are also often highly conserved. Thus, these positions can be assigned with moderate accuracy to the active site.

An alternative algorithm for assigning residues to the active site involves looking for conserved strings of amino acids, stretches 3 or 4 positions long that are conserved in a subgroup with an MPI of 50%, or 4–6 positions long conserved in a subgroup with MPI = 60–70%.

Finally, active site residues can be assigned in ways special to each class of protein. For example, mammalians Adh's have a highly variable substrate specificity, while fungal Adh's do not. Some of the variability in substrate specificity in mammalian enzymes must be due to variation in residues in the substrate-binding region of the protein. Thus, an algorithm that searches for positions that are variable in mammalian Adh's with an MPI of 85%, but invariant in fungal Adh's with an MPI of 50% (i.e., variability splits), can be designed. In Adh, this algorithm successfully identifies every amino acid protruding into the substrate binding pocket.

These algorithms strongly assign approximately 60% of the positions in the alignment of Adh's either to the surface, the interior, or the active site of a protein. Another 25% of the positions are assigned by similar algorithms more weakly to these regions. These form the basis for an assignment of secondary structure.

In assigning secondary structure, the first step involves dividing the alignment into small pieces that correspond (approximately) to segments of secondary structure. This is done through the use of "parsing algorithms". In decreasing order of reliability, these algorithms divide the sequence at positions that include deletions or insertions, APC Pro or Gly, Pro distributed over 2–4 adjacent positions in the alignment, splits at MPI = 50–85% containing Pro, Gly, Ser, Asn, or Asp, and Pro incompletely conserved but present in more than 3 subgroups at MPI = 60% or higher.

Such parsing elements normally, but not infallibly, occur between secondary structures. As examples of exceptions in the Adh's, a small part of one beta strand is deleted, and deletions occur a full turn into one alpha helix. However, unlike the algorithms mentioned above, the value of parsing algorithms is not significantly diminished by their missing on some occasions a turn of a helix or a residue in a beta strand (these will be picked up later during refinement).

Once the alignment is parsed, secondary structural elements are assigned. The maximum length of a helical unit or a beta strand can be estimated assuming that a protein adopts a globular structure. For alcohol dehydrogenases, the maximum length of a helix is approximately 22 amino acids, of a beta strand, 11 amino acids.

To assign secondary structure, one begins at one end of the protein and proceeds to the first strong parsing element. If the segment is longer than the maximum length of the longest presumed secondary structural element, it must be broken into small segments using secondary parsing elements. Assigning secondary structures to the intervening segments requires examination of strings of the sequence and their properties. Helices and beta strands are often characterized by a higher proportion of conserved elements (especially splits at MPI = 50–80%) than turns and coils. Simple coils and turns often contain a single hydrophobic split. Beta strands often contain consecutive splits; alpha helices contain splits that are found with a periodicity of 3.6 amino acids (corresponding to the helical periodicity). Thus, if the internal parts of a parsed segment contain consecutive splits (2 or more in subgroup with MPI = 50%, 3 or more with an MPI of 60%, or 3/5 at MPI = 50 or 4/6 MPI = 60%), coils or turns are excluded, and beta strands or helical secondary structures are considered.

Alpha helices that lie on the surface of a globular protein can be directly and reliably recognized by a pattern of internal and external residues (assigned in step b) appearing with a periodicity of 3.6 amino acids. The periodicity can be most quickly identified by plotting the segment on a helical wheel [121]. The 10–15% of the residues within the segment that have no designation (internal or external) are generally ignored in this analysis. In several cases, they can be weakly assigned "hydrophobic" or "hydrophilic" based on the average polarity of the residues at the position, and these added to the helical projection to confirm the assignment. An example is shown in Fig. 8 for the segment separating parsing position 365 and position 379. The segment is readily assigned as a surface helix, and the assignment is correct. This algorithm is extremely reliable even with alignments containing as few as 4 sequences.

Periodicity of 2 residues identifies a beta strand. However, unlike with alpha helices, surface beta structures are rare, and an alternation of surface and interior residues is normally seen only at the end of strands. However, here it is necessary to return to the definitions given near the beginning of this section. These definitions constitute a "metalanguage", providing terms (e.g. "amphiphilic split", "hydrophobic variable") that are abstractions of the extent and nature of the variability at a position in an alignment. The abstraction is more managable than the sequence data themselves and, if correctly constructed, contain all of the relevant information needed to assign secondary structure without recourse to structural interpretation (inside or outside).

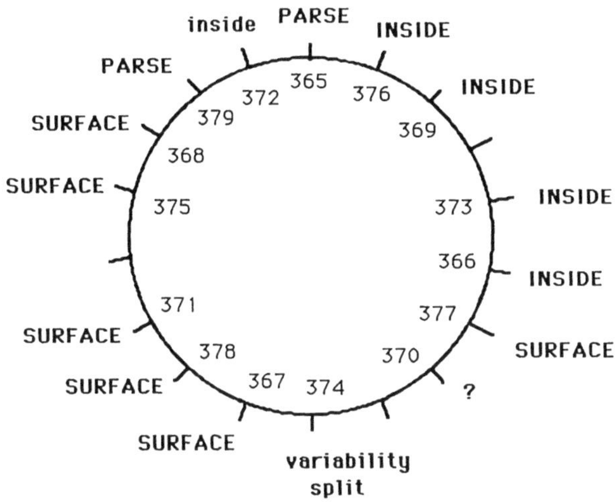

Fig. 8. A helical wheel of the segment of the alcohol dehydrogenase alignment between parsing positions 365 and 379 showing clearly the amphiphilicity of the helix

For example, the metalanguage assigns the abstraction "Interior" and "Amphiphilic split" alternatively to positions 36–42 (Table 5). We have not discussed (here) what the structural implications are for a position that is an "amphiphilic split with an MPI = 60%. But it is not necessary to know these implications for the 2-residue periodicity in this segment to be obvious, and to assign a beta structure to this segment based on this periodicity. This assignment is also correct.

Secondary structural assignments obtained by this approach are better than 80% accurate. Errors most often involve misidentification of beta strands and short coils. Nevertheless, in the best cases, the secondary structural assignments can serve as a starting point for constructing a tertiary structural model. In this construction, the elements must be oriented around a focal point. For enzymes, the active site is a convenient focal point, as evolutionary constraints on drift are quite special in this area.

Table 5. Alternating pattern of attributes used to identify a β STRAND

Position	Attribute
36	Interior
37	Amphiphilic split, MPI = 60%
38	Interior
39	Amphiphilic split, MPI = 60%
40	Interior
41	Amphiphilic split, MPI = 60%
42	Interior

As mentioned above, several algorithms can be used to assign positions at or near the active site. These assignments "tag" ends of secondary structural units, marking them as either pointing towards the active site or away from the active site. With Adh, the constraints on folding are fortunately sufficient to permit to identify the protein as a representative of a particular taxonomical class of protein as illustrated by Richardson, Rossmann, and others. This model is then confirmed or modified by analysis of variable residues that are likely to lie internally (vide supra), preparing a structure that can be the starting point for computational refinement of structure.

Those familiar with the literature on the prediction of tertiary structure may be disappointed at the complexity of this approach. There is a widespread sentiment that the solution to the protein folding problem, when it comes, will be a folding "code", perhaps in the form of a distributable computer program that can be used routinely by the biochemist whose primary expertise might lie somewhere else, and be equally applicable to membrane proteins, soluble proteins, binding proteins, structural proteins, and enzymes. This sentiment in part derives from the enormous ease with which a non-expert can use computer programs that use Chou-Fasman, hydrophobicity, or other classical algorithms [120].

Here, an evolutionary view is especially valuable as it constrains the bioorganic chemists to a reasonable set of expectations. Because natural proteins desire conformational instability, it is unlikely that the folded form of a protein will ever be determined from a single sequence by an abstract set of rules. At best, one can hope for a structural prediction only using a direct and complete computational method for finding a small energy minimum in a large conformational space.

However, once evolutionary constraints are accepted, they suggest avenues where progress can be made. The method reviewed above can provide a low resolution structure of Adh from fewer than 17 sequences. For a set of proteins as complex as Adh (the fact that some proteins are dimers and some are tetramers, and some have different functions than others make this a particularly complicated case), the goal now remains to improve the algorithms so that an equally satisfactory structural model can be obtained with a smaller number of sequences. This improvement too will come only after a better understanding of the evolutionary constraints on the divergence of protein sequences.

4 History, the RNA World, and Early Life

Adaptation and drift are not the only evolutionary processes that determine chemical behavior in living systems. As noted in the introduction, some non-adaptive traits do not drift simply because there is no mechanism that permits drift without disrupting other selected traits. Such constraints on drift imply that some macromolecular behaviors will be vestiges of non-functional behaviors of ancient macromolecules.

Vestigial traits are well known in biology, and many physiological details in higher organisms can best be understood as remnants of traits in more ancient organisms [122]. Vestigial traits are expected in biological macromolecules if one of two (or more) alternative chemical solutions to a biochemical problem must be chosen, if either alternative is equally suited from the point of view of survival, and if the choice is difficult to alter once it is made. For example, it should make little difference which amino acid is coded for by the nucleoside triplet GGG in the genetic code. However, the triplet code must for a single amino acid. Further, once a choice is made, the codon will be used in genes for many proteins, and changing the meaning of the codon would require codons in all of these genes to be changed simultaneously. Thus, the modern code almost certainly contains vestiges of very early codes.

Similarly, the use of L-amino acids in proteins, D-sugars in nucleic acids, and ribonucleotide fragments in cofactors (such as ATP, NAD^+, FAD, S-adenosylmethionine) reflect no obvious selected function, but are difficult to alter without a major revision of metabolism. Although functional models for these aspects of modern biochemistry have been offered [10, 123], they most likely reflect random choices early in the history of life on earth that have been conserved.

Thus, a historical model describing events in the evolution of modern biological macromolecules is an essential part of any effort to unify evolutionary theory and structural theory. Historical models are ad hoc; they must postulate random events in addition to events that are predictable from physical laws. Organic chemists, who deal in the properties of molecules that are immutable in time, are often uncomfortable with such models. Nevertheless, historical models play important roles in many sciences, including geology (where the present state of the earth is described as the result of past movements of plates on the earth's crust) and linguistics (where the ancient proto-Indo-European language is reconstructed by extrapolation from the structures of modern languages).

4.1 The Rules of the Game

Historical models attempt to reconstruct the biochemistry of ancient forms of life by an extrapolation of the details of modern biochemistry, in particular, the structures, sequences, and behaviors of biological macromolecules. As in paleontology, where entire organisms are reconstructed from fragments of bones, substantial reconstruction must be done from limited information. Model building begins with definitions of rules for extrapolation and reconstruction. A successful model is one that is internally consistent, is chemically plausible, and explains a wide range of modern biochemical traits in an intricately interconnected conceptual framework that allows the design of experiments.

To model ancient macromolecules, one must classify the organisms that are the sources of modern macromolecules. We use here a scheme that divides

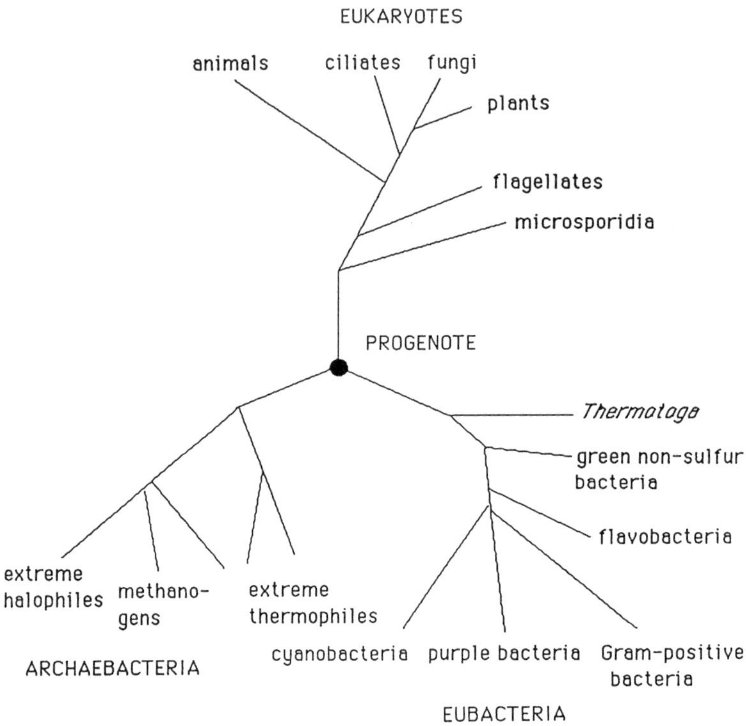

Fig. 9. An evolutionary tree showing the major descendants of the progenote, archaebacteria, eubacteria, and eukaryotes

modern organisms into three kingdoms, archaebacteria, eubacteria, and eukaryotes (Fig. 9). At the molecular level, archaebacteria and eubacteria are as different from each other as each is from eukaryotes, and these three kingdoms can be viewed as independent lineages descendant from a common ancestor, which we shall call the "progenote". This three-kingdom classification is the starting point for comparing modern biochemical data to reconstruct ancient forms of life. Alternative classifications change our discussion only subtly [124].

Working Backwards: The Rule of Parsimony

A detailed model of the progenote, the most recent common ancestors of all forms of life in these three kingdoms, is the first goal of the reconstruction. The biochemistry of the progenote is reconstructed after examination of the biochemical similarities and differences in its descendants using a rule of "parsimony" [125]. A parsimonious model for an ancestral protein is one where the biochemical diversity in its descendants can be explained by assuming the minimum number of independent evolutionary events. Parsimony can be directly used to propose a hypothetical sequence for an ancestral protein when

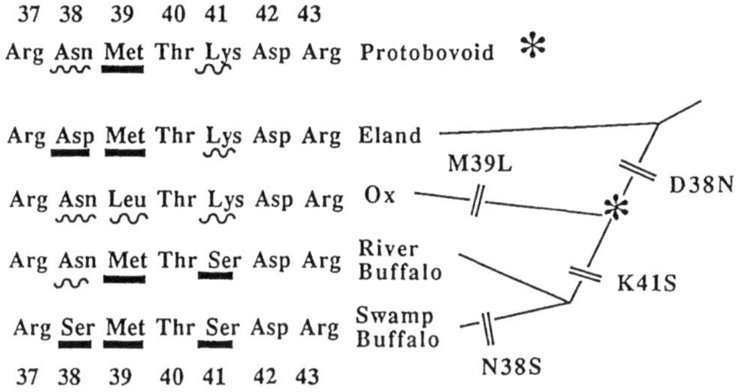

Fig. 10. An illustration of the use of parsimony to construct part of the sequence of the ribonuclease present in an extinct organism, "proto-bovoid," the most recent common ancestor of ox, river buffalo, and swamp buffalo. The notation on the tree (right) uses the one letter code for amino acids

the sequences of at least 3 of its descendants are known. Application of parsimony in such a case is illustrated in Fig. 10 for ribonucleases.

Similarly, metabolic pathways of ancient organisms can be reconstructed to explain modern metabolic diversity with the smallest number of postulates concerning arbitrary historical events. If enzymes catalyzing several steps in a metabolic pathway are homologous in descendants, the ancestor probably contained this pathway. Likewise, the distribution of natural products can implicate the presence of pathways for the biosynthesis of these natural products in ancestral organisms [126].

An evolutionary tree describing the relationship between modern organisms can be constructed using parsimony. The tree should, in principle, be superimposable on a tree describing the pedigree of the organisms containing these enzymes, which themselves are deduced from independent data, such as the fossil record. In practice, discrepancies are observed from time to time between trees derived from examining a single protein, and those derived from an overview of multiple physiological and biochemical traits. For example, in the tree derived from RNase sequences by Beintema [127], the giraffe and pronghorn antelope might be too closely related, judging from an analogous tree derived from the fossil record (Fig. 11). Such discrepancies may simply reflect the statistics of data collection, and reflect the well-known fact that evolutionary trees are best constructed from joint consideration of many independent molecular and physiological determinants. However, such discrepancies might also reflect convergent evolution, an incorrect fossil tree, lateral transfer, or other processes where genetic information moves between organisms in ways other than by lineal descent (vide infra).

Obviously, the more descendants of a particular ancestor that are available for inspection, the easier it is to obtain a reliable picture of ancestral proteins and metabolism. With only two descendants, parsimony only allows the hypothesis that traits shared by the descendants were present in the common

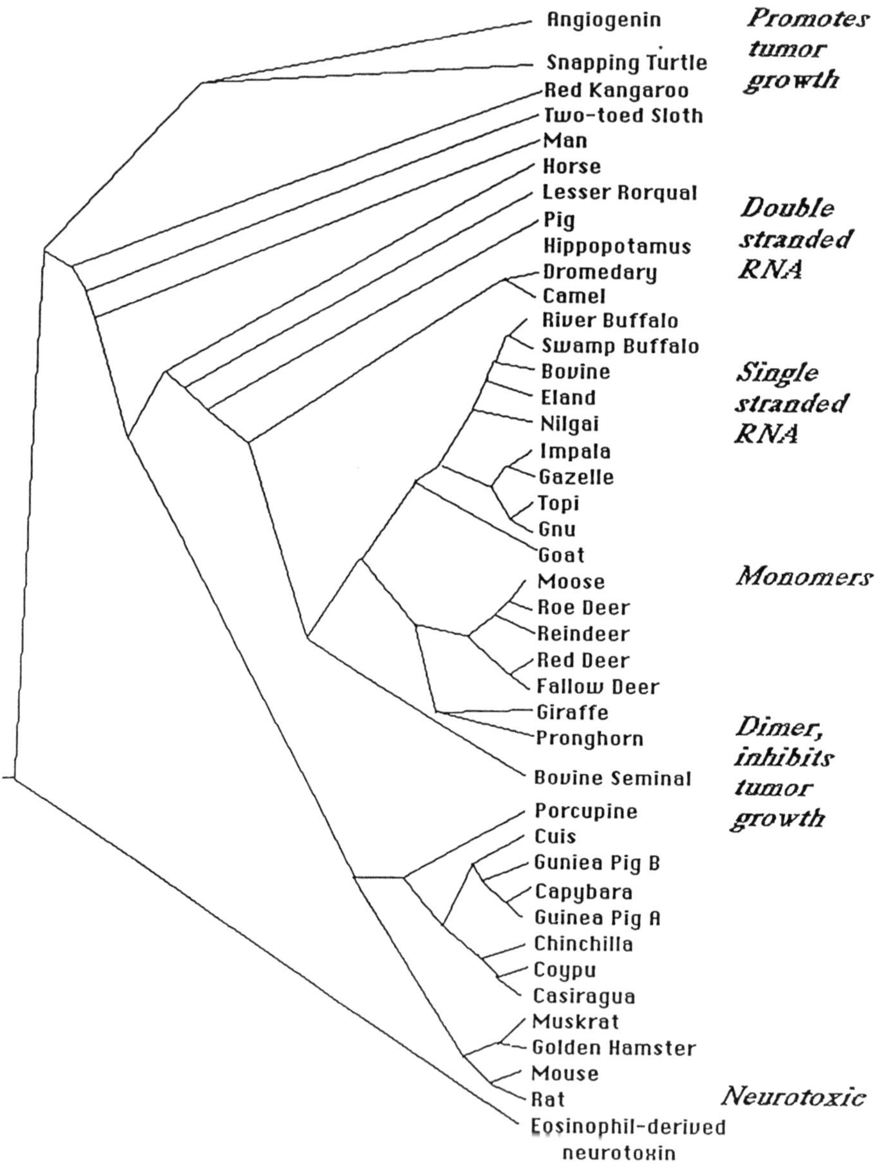

Fig. 11. An evolutionary tree for ribonucleases and their homologs, angiogenin and eosinophile-derived neurotoxin, showing some of the physical, catalytic, and biological divergence within this class of homologous proteins. After Ref. 127

ancestor. With three descendants, the rule of parsimony can be applied with varying degrees of stringency. Traits found in all three lineages can be securely assigned to the ancestor. Traits found in only two might also be assigned to the ancestor, but such assignments are clearly less secure.

Rules of parsimony come in various forms [128, 129], and choosing among them requires a sophisticated discussion that is inappropriate here, and unnecessary to understand, in general terms, how sequence, structural, physical and chemical data can be used to reconstruct pictures of ancient organisms. It is important to recognize that these constructs are models. As such, they are neither "true" nor "false". Rather, they are working hypotheses that rationally organize available data from modern organisms, predict traits in modern organisms, and therefore serve as the basis for experimentation. Aside from the complexity of the computation when it involves many sequences, the process is straightforward. However, three evolutionary events disrupt the simple application of parsimony in this reconstruction: convergent evolution, lateral transfer, and deletion–replacement events.

Convergent Evolution

If a trait serves as a selectable function, it can arise independently in separate lineages. Wings are found in both birds and bees, but it is a clear error to conclude from parsimony that the most recent common ancestor of the two organisms could fly. The error is obvious given an evolutionary biologist's understanding of morphology in these anatomical structures. Both sets of wings are adaptive, and could have arisen independently through natural selection on separate lineages for their survival value. Further, the structures of the two sets of wings are so clearly different that they cannot be homologous; one cannot imagine evolution of one to the other by a continuous process.

Similar cases are conceivable in molecular evolution. If a macromolecular trait is a unique chemical solution for a particular biochemical problem, it might have arisen independently in the three kingdoms of life (convergent evolution), and not have been present in the most recent common ancestor. Thus, special caution must be used in assigning functional traits that are supported by very little sequence information. For example, noting that enzymes catalyzing the reduction of sulfur were present in some members of all branches of his tree, Lake concluded that the progenote was a sulfur metabolizer [124]. This may be true. However, a purely chemical perspective suggests that enzymes can easily arise to reduce sulfur (vide supra), and such enzymes should be strongly adaptive in particular environments. Thus, such an assignment should be considered weak until it is supported by sequence data.

Traits judged not to reflect a unique chemical solution to a particular biochemical problem, if found in all three kingdoms, are more reliably assigned to the progenote than traits that do. The organic chemist's understanding of chemical reactivity in biological macromolecules is necessary to make this judgment. However, sequence data can be decisive. For example, nearly identical pathways for purine biosynthesis and histidine biosynthesis are found in archaebacteria, eubacteria, and eukaryotes. The organic chemist intuitively views the chemistry of the pathways to be inelegant; it is certainly not a chemically unique solution to these biosynthetic problems and probably not an optimal one. On this basis alone, the two pathways might be assigned to the progenote. However, the sequences of several enzymes catalyzing steps in the pathway are known from members of the three kingdoms; several of the enzymes are clearly homologous [130, 131]. Thus, the progenote almost certainly biosynthesized purines and histidine by essentially the same pathways used in modern organisms.

However, sequence data do not necessarily provide absolutely unambiguous determinants of homology. Although it is true (vide supra) that many enzymes with quite different sequences can perform the same function, it is conceivable that some sequences are uniquely suited to perform a particular biological function. As noted above, the sequences of lysozymes may have converged to reflect convergent environment in "ruminant" primates and ungulates. Examples of "sequence convergence" are extraordinarily difficult to document and seem to be rare. Nevertheless, the possibility remains that evolutionary trees based on a single determinant will, from time to time, be incorrectly constructed as a result.

Lateral Transfer

Parsimony works only if the only mechanism for transferring genetic information between organisms is via "vertical" descent, from parents to children. If divergent organisms exchange genetic information after they have diverged ("lateral transfer" of genetic information), a rule of parsimony will incorrectly place these organisms closer together on an evolutionary tree than appropriate. The possibility of lateral transfer also introduces uncertainty into assignments of primitive traits based on a comparison of descendant organisms.

Lateral transfer can be proposed whenever an evolutionary tree constructed for a particular protein differs significantly from that constructed using other determinants. Lateral transfer of genetic information between organisms is well known, with a frequency that depends on the relationship between the donor and acceptor organisms. The exchange of information between viruses and their hosts is so facile that viruses and their hosts are best considered to be a single organism. For example, sequences of thymidylate synthase (2.1.1.45) from the herpes virus is 70% identical to that from humans [132]. The DNA polymerases (2.7.7.7) of man and animal viruses also appear closely homologous. This does not mean that all viral genes are closely homologous to analogous genes from their host; for example, exonucleases (3.1.11.1) from *E. coli* and two of its viruses, lambda and T7, appear to be non-homologous [133]. However, evolutionary trees for viruses cannot be constructed by parsimony independent of the trees for their hosts. Further, to the extent that viruses and other non-chromosomal elements of DNA (e.g. plasmids and transposons) can cross species boundaries, they offer mechanisms for the lateral transfer of genetic information between species. Likewise, genetic information has probably been transferred between the nuclear genomes of eukaryotes and the genomes of eukaryotic organelles (mitochondria and chloroplasts) that presumably are eubacterial in origin [134–136].

Lateral transfer is also well known between different species of eubacteria, and is the basis for the transfer of drug resistance between bacteria [137]. In higher organisms, the sequences and distribution of introns and other non-coding DNA elements suggests in several cases that lateral transfer has occurred between species within a kingdom [138–140]. This suggests that these non-coding elements are not good determinants for constructing evolutionary trees.

This conclusion is especially significant in light of recent molecular biological literature which suggests that intron position is a reliable evolutionary marker [141]. While this appears to be true in some cases [142], it is clearly not true generally. For example, none of the 8 introns in the mammalian gene for alcohol dehydrogenase coincide in position with the 9 introns in the highly homologous (50% at the level of protein sequence) genes from plants [143, 144]. Introns are also problematic evolutionary markers in genes for actin [145], fibrinogen [144], cytochrome P450 [147, 148], and Ras oncogenes [149].

Eukaryotic genes coding for proteins may also have been transferred laterally between members of a single kingdom. For example, "silent" variation (vide supra) in histone genes from two widely divergent sea urchins is far less than expected given the time since the organism's divergence. Either a special selective force constraining silent drift must be assumed, or the gene has been transferred between species subsequent to their divergence [150]. Lateral transfer of genetic information between parasites and their hosts and between symbionts also has been proposed. For example, the hypoxanthine–guanine phosphoribosyltransferases (2.4.2.8) from the malarial parasite *Plasmodium falciparum* show extensive homology (48%) with the enzyme from mouse, again suggesting lateral transfer [151]. More sequence data are needed to determine the degree of sequence homology between enzymes from these organisms generally to evaluate the plausibility of this suggestion.

Lateral transfer may also occur between kingdoms. For example, the sequence of the glutamine synthetase II (2.6.1.15) gene from the eubacterium *Bradyrhizobium japonicum*, which lives in close contact with plants, is more similar to the gene for the analogous enzyme from the plant than it is with other eubacteria, strongly suggesting lateral transfer from eukaryotes to eubacteria [152, 153]. Likewise, genes for isopentenyltransferases (2.5.1.27), enzymes involved in the formation of cytokinins produced by plants and by many microorganisms living in contact with plants, are clearly transferred from bacteria to plant cells during tumor induction, and may have originated in the two kingdoms by lateral transfer [154]. Lateral transfer of genes for copper/zinc superoxide dismutase (1.15.1.1) between the ponyfish and its symbiont, *Photobacterium leiognathi* has also been suggested [155, 156]. However, in light the sequence similarities of superoxide mutases generally, the similarities between the eubacterial and eukaryotic enzymes may be explained more simply in terms of vertical descent [157].

Deletion–Replacement Events

Deletion of a gene for a function followed by the re-emergence of the function by a cross-evolution of a replacement macromolecule (vide supra) is a general mechanism that can create the appearance of rapid divergence in proteins, and its effects on our view of molecular evolution are under-appreciated. Deletion–replacement processes complicate the superimposition of a tree derived from sequences with the true divergent evolution of the organisms themselves, and tend to diminish the total number of independent lineages represented in a single organism. Indeed, given sufficient evolutionary time in an organism that is not creating new proteins de novo, the process will eventually produce organisms where all proteins are homologous.

This process produces results that can easily be misinterpreted. For example, Richardson [158] has noted that a limited number of folded forms are found in modern proteins. Many have inferred from this observation that there exist only a limited number of folded structures are stable, or that all proteins in a particular taxonomic class are descendants of a single primordial protein of this class present in the progenote [158, 159]. However, the observation can also be explained by randomly occurring deletion–replacement events in recent organisms, processes that would gradually eliminate much of the diversity in protein structures found in ancestral organisms.

The frequency of deletion–replacement events should be different for different types of proteins and in different organisms. The most common deletion–replacement events should affect proteins participating in the biosynthesis of metabolites that can be obtained in the diet. Deletion–replacement events should also be frequent when an organism already contains proteins that perform analogous functions, as such proteins can assume the function of the deleted protein after only small structural alteration. Proteins that bind the same cofactor, product or substrate of the deleted protein (including proteins catalyzing the formation of the substrate or subsequent reaction of the product) are all candidates. In contrast, if the deletion is lethal under all environments, or if proteins performing chemically analogous functions are not available in the organism, deletion–replacement events should be rare.

For example, replacement of a ribosomal subunit by a deletion–replacement event is unlikely: the deficiency arising from the deletion cannot be easily compensated by diet, and no protein in normal metabolism performs a function closely analogous to message-directed biosynthesis of proteins that is available as a rapid replacement. However, beta-galactosidase (vide supra) should undergo frequent deletion–replacement events. The deletion is selectively disadvantageous only in special environments, and an organism with a complex metabolism contains many proteins that catalyze glycosyl transfer reactions that are chemically similar to those catalyzed by beta-galactosidase. Each of these could be made into a replacement galactosidase by only a few point mutations.

The RNA World

The similarities displayed in the three kingdoms make inescapable the conclusion that all of modern life arose from a single organism, called the "progenote", the most recent common ancestor of archaebacteria, eubacteria, and eukaryotes. A biochemical trait can be assigned to the progenote most strongly when: a) the trait is found in several representative organisms from each of the three kingdoms; b) assignments of homology in various branches of the progenotic pedigree are supported by high information content (preferably sequence data); and c) the trait does not represent a chemically unique solution to a particular biochemical problem in the modern world. Such assignments are not absolute; if only some of these criteria are fulfilled, a weaker assignment can be proposed.

However, it is increasingly clear that the progenote was not the first organism. Indeed, to explain many of the biochemical peculiarities mentioned above, we must assume that life on earth passed through at least three episodes (Fig. 12) [15, 160–170]. In the first (the "RNA world" [162]), catalytic RNA

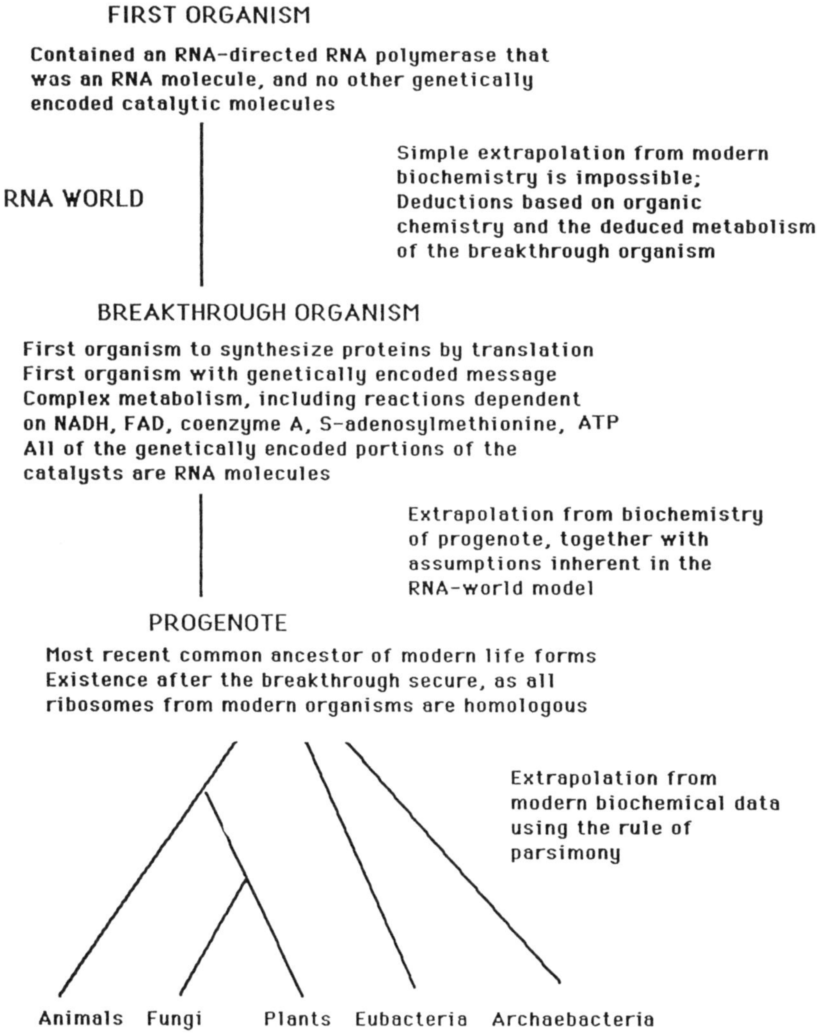

Fig. 12. An overview of the three episodes of life on earth for which biochemical evidence remains or can be inferred

molecules ("riboenzymes") were the only genetically encoded components of biological catalysts.

The beginning of the second episode was marked by a "breakthrough" to translation, the first synthesis of a protein by the translation of a genetically encoded RNA message in a "breakthrough organism" [15,160,161]. The breakthrough organism is also the first organism to contain a functioning ribosome, which (in this model) consisted of RNA as the sole genetically encoded component. Following the breakthrough, both proteins and RNA molecules served catalytic roles (as they do in the modern world).

Many components of modern ribosomes are homologous in all three kingdoms. Assuming that vertical transmission of genetic information is the rule, this means that the progenote must have lived after (or contemporaneously) the breakthrough. Thus, the progenote, despite its name, represents only the beginning of the most recent episodes in the history of life.

Rich [280], Woese [164], Crick [165], and Orgel [166] in the 1960s noted that such a model is consistent with the structure of t-RNA. Further, the model solves or explains four important problems or facts: a) the "chicken or egg" problem, originating in the simultaneous need for DNA to make proteins, and proteins to make DNA; b) the intermediacy of m-RNA between DNA and proteins. c) The presence of r-RNA as the principal component of ribosomes; d) The fact that many cofactors contain RNA-like moieties that do not participate in enzymatic reactions (Fig. 13). The model was developed in the 1970s, and can be found in textbooks of the era [167]. In 1976, Usher and his coworkers discovered the first (although poor) example of catalytic RNA [168]. Nevertheless, RNA was considered to be too poor a catalyst for the RNA world to have lasted for long. Catalysis by RNA molecules was reported as early as 1961 [281].

The full significance of the structure of many "ribo-cofactors", first mentioned by Orgel, was appreciated and developed by White [169]. Finally, in what might be regarded as the apogee of theoretical development of this model to date, Visser and Kellogg provided the first clear interpretation of the reactivity and structure of various cofactors in terms of this model [170]. In the 1980s the model became popular with molecular biologists following the surprising discovery of self-splicing messenger RNA molecules and catalytically active RNA involved in the processing of transfer RNA [171–175].

Modeling the Last Riboorganism

As vestiges of the RNA world apparently can be found in modern biochemistry, it too must be considered in a complete model that combines evolutionary and structural theory. However, parsimony cannot be used to determine which traits in the progenote are vestiges of the breakthrough organism. The breakthrough organism apparently left only one descendant, the progenote, for which we have an (incomplete) model reconstructed from the biochemistry of *its* descendants, and parsimony works only when one can compare the biochemistry of many descendants of an ancient organism.

Thus, to the extent that a metabolism for the progenote can be constructed, *chemical* criteria must be used to identify which details of this metabolism are

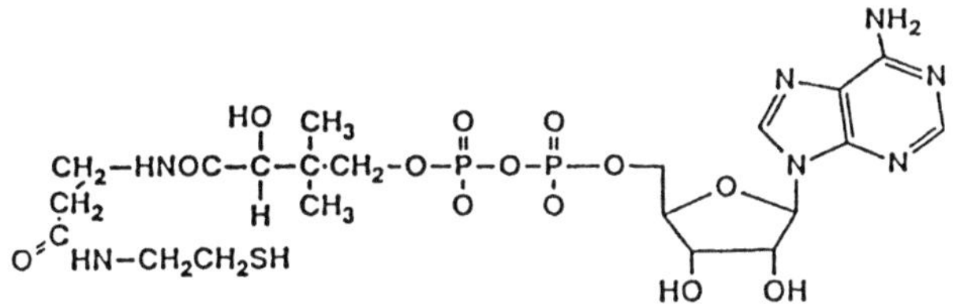

Coenzyme A

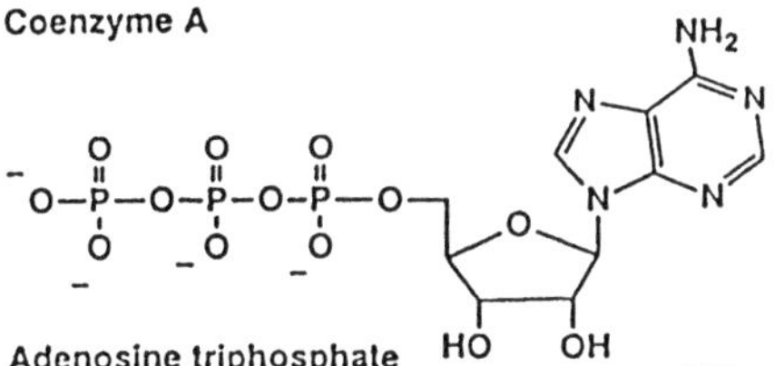

Adenosine triphosphate

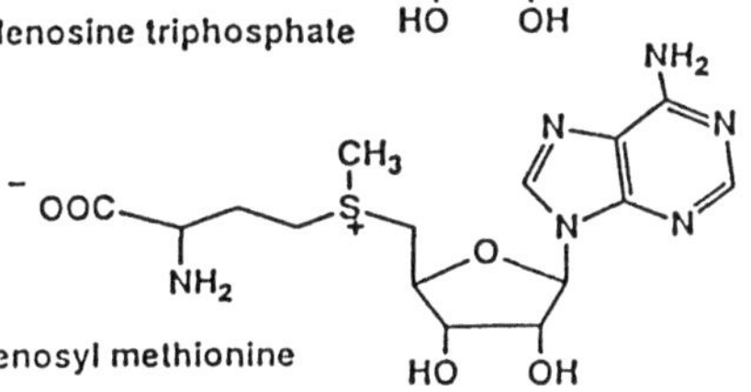

S-adenosyl methionine

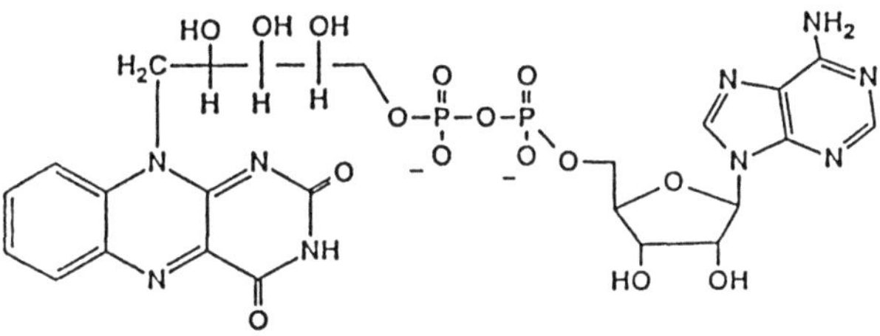

Flavin adenine dinucleotide

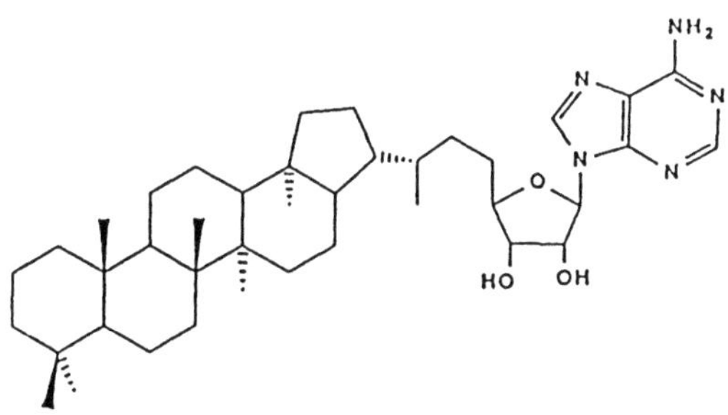

30-(5'-adenosyl)hopane Rhodopseudomonas acidophila

Fig. 13. Structures of four "ribo-cofactors", molecules that incorporate fragments of RNA that are not directly involved in the chemistry of the cofactor, and one "ribo-terpenoid" from *Rhodopseudomonas*. These structures are presumably vestiges of the RNA world

vestiges of the RNA world. These rules are based on the assumption that RNA molecules present in the progenote nearly always emerged in the RNA world. A biochemical trait of the progenote can be assigned to the breakthrough organism most strongly when a) RNA is involved in the trait, b) the involvement does not reflect the intrinsic chemistry of RNA, and c) substitution of another structural unit for the RNA unit could, on chemical grounds, provide similar or better biochemical performance.

Using these rules, ribosomal RNA, already reliably placed in the progenote, can be placed in the breakthrough organism. Likewise, RNA cofactors (NAD^+, SAMe, CoA, ATP, FAD) all contain fragments of RNA (Fig. 13) that are present in all of the lineages descendant from the progenote, and such ribocofactor structures can be assigned to the progenote. The RNA portions of the cofactors are not essential for their chemical roles in modern metabolism [176, 177]. Thus, on chemical grounds, these ribocofactors can be assigned to the breakthrough organism, together with ribo-enzymes that used these cofactors. This assignment requires the assumption that the breakthrough organism was metabolically complex, with riboenzymes that catalyzed redox reactions, transmethylations, carbon–carbon bond formation, an energy metabolism based on phosphate anhydrides, and carbon–carbon bond forming and breaking reactions.

A similar argument can be made for RNase P, a ribonucleoprotein in both eukaryotes and prokaryotes that catalyzes the biosynthesis of t-RNA molecules [172]. The RNA portion can be assigned with moderate conviction to the progenote (information is not yet available for archaebacteria). As the RNA portion again serves no function that intrinsically must be performed by RNA, its origin is assigned to the RNA world. This assignment implies that certain aspects of modern organization of t-RNA genes are also vestiges of the RNA world. These hypotheses could be evaluated by a study of the RNase P from archaebacteria.

Conversely, biotin and *S,S*-dimethylthioacetate display chemistry expected for cofactors that arose in the protein world. This point, eloquently stated 10 years ago by Visser and Kellogg [170], remains unappreciated by several groups building models of the RNA world [178].

The "nearly always" qualification in the rule that "RNA molecules present in the progenote nearly always emerged in the RNA world" is required because in some cases, the participation of RNA in a biochemical process might be a unique chemical solution to a particular metabolic problem. In such cases, participation of RNA is expected to arise even in organisms where proteins play a major catalytic role. Here, as before, modern metabolic processes that exploit the intrinsic chemistry of a biological macromolecule are not good evolutionary markers.

For example, consider the proposition that ubiquitin is a vestige of the RNA world. A proponent would note that ubiquitin is present in all eukaryotes, allowing it to be assigned to the "proto-eukaryote" [179]. However, ubiquitin is not known to be present in prokaryotes or archaebacteria, making an assignment to the progenote unreliable. However, the advocate might argue that insufficient effort has been made to find ubiquitin in archaebacteria and eubacteria, so we cannot *exclude* ubiquitin from the progenote.

Next, the proponent might point out that uncharged t-RNA molecules are essential components of the ubiquitin-dependent proteolytic system [180]. He would interpret this as evidence that ubiquitin should be assigned to the RNA world. This assignment would be indentifiably weak, as the activation of ubiquitin by uncharged t-RNA molecules makes good metabolic sense: uncharged t-RNA implies a deficiency of amino acids, for which degradation of proteins via the ubiquitin-dependent process appears to be a biologically adapted remedy. Thus, while t-RNA plausibly arose in the RNA world, the involvement of t-RNA in ubiquitin-dependent degradation of proteins appears to be functional, and would be expected to emerge for functional reasons, even in organisms that already employ proteins as the principal catalytic molecule.

Self-splicing introns, 3′-tRNA structures, and certain non-coding details of eukaryotic genes [174], similarly cannot be reliably assigned as vestiges of the RNA world. They are not known (at present) in all kingdoms, meaning that the rule of parsimony cannot reliably place them in the progenote. Several of the structures have apparently been transferred laterally (vide supra). In several cases, strong arguments can be made that the traits are unique chemical solutions to biochemical problems, and could have evolved in a world of protein catalysts. For example, self-splicing introns can be viewed as a chemically sensible adaptation where introns must be removed from messages. They exploit the intrinsic reactivity of RNA and the relative facility of catalyzing transesterifications at phosphorus. Further, self-splicing appears to be simpler than alternative splicing mechanisms that involve ribonucleoproteins. Indeed, splicing processes that are *not* self-catalysed are more likely to be vestiges of the RNA world [181].

The progenote is expected to combine both traits left over from the RNA world with those that emerged after the breakthrough. Most conspicuous of the new traits are protein catalysts. For pathways that emerged after the breakthrough, all of the enzymes should be proteins except in rare cases where the intrinsic chemistry of RNA makes it preferable as a catalyst. However, in the time separating the breakthrough from the progenote, proteins should have arisen by deletion–replacement processes to replace at least some of the riboenzymes that catalyzed steps in pathways originating in the RNA world. Still more deletion–replacement events may have occurred following the divergence of the progenote. In these cases, the proteins in the modern three kingdoms might well be non-homologous.

Prebiotic Chemistry

Studies of "prebiotic" chemistry complement the rule of parsimony to modern biochemical traits. For example, recent work of Eschenmoser and his coworkers has revolutionized our view of the corrin ring of vitamin B-12; instead of being a "complex" capable of synthesis only by living systems, corrins might actually be products of prebiotic chemistry [182]. Likewise, experiments pioneered by Miller, Ferris, and others have provided much insight into the molecules that might have been formed abiotically, and therefore available to early forms of life [183].

Origin of New Catalytic Functions

Models of the RNA world also must consider the relative facility of evolving new catalytic functions. Our understanding of the evolution of new catalytic function in the modern world suggests three important issues: a) How much information must be assembled in a biological macromolecule before it can perform a selectable function? b) Could this information come from other biological macromolecules that have previously evolved to serve another function? and c) What is the selective force demanding the new metabolic process?

The first question requires the intuition of the organic chemist to answer, and is based on a subjective evaluation of how easy is it to catalyze a particular reaction. For example, catalysis of the hydrolysis of RNA, the decarboxylation of beta-ketoacids, the transfer of phosphoryl groups, and the hydrolysis of esters is "easy" compared to catalysis of the hydrolysis of amides, the decarboxylation of alpha keto-acids, and the transfer of methyl groups [10].

Metabolic steps that are easy to catalyze should arise faster than those that are more complex. For example, the chemical "information" that must be assembled in a ribosome before it can positively contribute to survival via an ability to synthesize proteins is considerably larger than that for conceivable RNA's that would catalyze most metabolic steps (or entire pathways). The direct implication is that metabolism preceded translation, and that the ribosome in the breakthrough organism arose by a combination of metabolic riboenzymes that served other selected roles in ribo-metabolism.

Indeed, translation machinery appears to be so complex that individual components of the ribosome must have performed other roles before adopting a role in translation. Non-translational synthesis of oligopeptides might well have been selected before the breakthrough, and the amino acid-RNA conjugates involved in cell wall biosynthesis and chlorophyll biosynthesis could be vestiges of these processes in the RNA world [15, 16].

Arguments from Biology

The ecological behavior of metabolically complex organisms living in an RNA world should be similar to that of organisms in the modern world. For example, the existence of ribo-autotrophs implies the existence of ribo-heterotrophs. Modern organisms that specialize to adapt to ecological niches have a survival advantage; similar advantages are expected to accrue to riboorganisms. Adaptation involving speciation and genetic isolation of organisms in the modern world is also expected in the RNA world. Likewise, competition for resources between organisms produces extinction in the modern world; this is also expected in the RNA world.

Thus, if metabolic complexity emerged before translation, an ecologically complex RNA world should also have developed. The breakthrough must have conferred selectable advantage on the organism possessing it. Therefore, following the breakthrough, there must have been extinction of non-translating organisms overlapping the environment.

4.2 Historical Conclusions from Sequence Data

We now review some general conclusions regarding historical views of the evolution of modern biological macromolecules, conclusions made possible by available sequence data. These conclusions permit an elaboration of a historical model for the origin of modern life in some detail.

Enzymes Catalyzing Identical Reactions in Different Organisms Are Generally Homologous By Sequence

The most obvious pattern apparent from sequence data is that enzymes catalyzing the same reaction in different organisms are generally homologous. This is certainly true within single kingdoms. Between kingdoms, it is largely true between eukaryotes and prokaryotes in central metabolic enzymes; the glycolytic pathway and the citric acid cycle are two of many examples.

Only the absence of sequence data from archaebacteria prevents us from evaluating this general statement across all three kingdoms. However, these data are now emerging. In purine and histidine biosynthesis [130, 131], enzymes from all three kingdoms are now known to be homologous. This suggests that the pathways themselves are homologous. Only slightly weaker cases can be made for DNA-directed RNA polymerases and certain other central enzymes in metabolisms.

Of course, ribosomal RNA is homologous in all three kingdoms; indeed, comparisons of homologous ribosomal RNA sequences is the basis for the three-kingdom model. However, several ribosomal proteins similarly are also homologous in the three kingdoms, and therefore can be assigned to the progenote [184]. Thus, the progenote contained a ribosome with protein subunits, a purine biosynthetic pathway, and a histidine biosynthetic pathway.

Either the pathways arose after the breakthrough (in which case, the only catalysts for this pathway were proteinaceous), or before the breakthrough (in which case most of the riboenzymes involved in the pathway had been replaced by proteins before the emergence of the progenote). Several arguments suggest the second hypothesis. First, the pathways avoid cofactors that most probably arose after the breakthrough. For example, the carboxylation step in purine biosynthesis does not involve biotin, which (in this model) originated after the invention of translation (Fig. 14). Further, ATP is used in the formation of the aromatic imidazole ring in purines. This use is chemically interesting, as ATP is not needed in this step to provide a thermodynamic driving force for the reaction. The use of ATP in this reaction (perhaps to overcome a kinetic barrier) appears wasteful, and might be expected in metabolic pathways originating in the RNA world that had unsophisticated ribo-enzymes as catalysts.

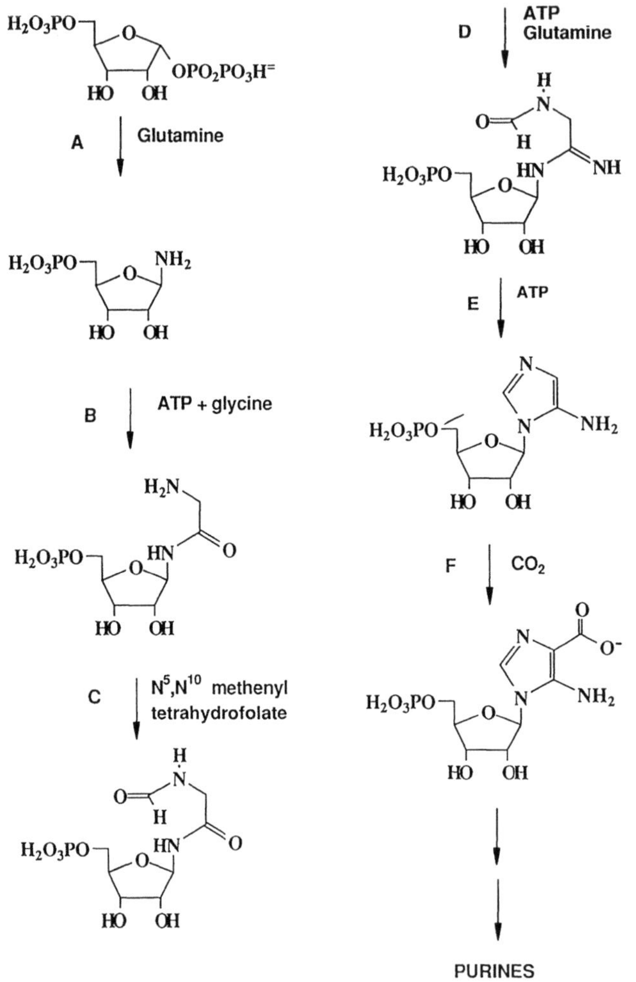

Fig. 14. The pathway for the biosynthesis of purines shows several characteristics of a pathway emerging in the RNA world. Notably, step F does not involve biotin, even though many other carboxylations do. In step E an aromatic ring is formed, a reaction that is expected to be thermodynamically spontaneous. Nevertheless, ATP is hydrolyzed, perhaps to overcome a kinetic barrier, suggestive of a primitive catalyst

Further, biological considerations suggest that as a central component of a riboorganism, biosynthetic routes to purines should have arisen early. Further, following the breakthrough, purines should have been available in the diet, suggesting that deletion–replacement events would have been facile, and the superiority of protein catalysts over riboenzymes suggests that most of the riboenzymes would have been replaced by protein enzymes soon after the breakthrough.

Purine biosynthesis and histidine biosynthesis are therefore good "paradigms" of a ribo-pathway whose ribo-enzymes were replaced by protein enzymes before the progenote. It is a benchmark against which other pathways can be discussed.

However, in many Cases, Different Solutions to Identical Mechanistic Problems have Arisen, Often in Enzymes Catalyzing Reactions that are not on Major Metabolic Pathways

It is common in secondary metabolism for essentially the same reaction to be catalyzed by two quite different enzymes in different organisms. In these cases, it is simplest to conclude that the enzymes were separately invented. Serine proteases provide the classical example of convergent evolution. Subtilisin (a eubacterial protease) and trypsin (a eukaryotic protease) show nearly identical mechanisms but quite different tertiary structures. As with the bird and bee wings, it is difficult to imagine continuous evolution from one to the other, and this is a strong argument that the two proteases are not homologous.

Several other examples are now available. For example, bacterial ribonucleases are not obviously homologous to their eukaryotic counterparts [185]. Enzymes catalyzing the hydrolysis of beta-galactosides (3.2.1.23) apparently have emerged more than once even within the eubacterial kingdom [186]. Cephalosporin acylase (3.1.1.41) from the eubacteria *Pseudomonas* has essentially no sequence homology with other acylases [187]. Examples such as these suggest either that a) the progenote did *not* have these enzymes, and the different enzymes originated independently in different lineages; b) the enzyme present in the progenote as a riboenzyme that was a vestige of the RNA world which was replaced by non-homologous proteins in all descendant lineages; or c) the progenote contained a proteinaceous RNase that was deleted and replaced in one line of descendants.

While non-homologous pairs of enzymes catalyzing identical reactions seem to be more common for enzymes acting in secondary metabolism, they are also known in enzymes of primary metabolism. For example, Kraut and his co-workers have reported that two bacterial dihydrofolate reductases (1.5.1.3) with no sequence homology have distinctly different folded forms, suggesting multiple inventions of this enzyme [188]. It is unlikely that the progenote lacked a pathway for the biosynthesis of thymidine, given the argument (vide supra) that the progenote biosynthesized purines and used DNA (vide infra). As the biosynthesis of thymidine requires derivatives of folic acid, the progenote most probably had a dihydrofolate reductase. Thus, the deletion–replacement events, or independent replacement of a ribo-enzyme in the progenote, are the most likely explanation for multiple pedigrees of this enzyme.

In some cases, a newly developed biological role may account for independent origins of a similar catalytic activity. For example, hydrogenase (1.18.99.1) is widely distributed in nature, and is found in many prokaryotes and in some eukaryotes (algae, green plants, protozoa). Several classes of hydrogenases are known, including those containing non-heme iron and those containing nickel.

However in the eubacterium *Desulfovibrio*, three types are known, one with non-heme iron, one with non-heme iron and nickel, and one with iron, nickel, and selenium. The *D. baculatus* NiFeSe enzyme shows no sequence similarity to the *D. vulgaris* Fe enzyme [189].

Multiple origins in this case suggest that the progenote did not contain a hydrogenase; indeed, in this case, one can argue that the most recent common ancestor of various strains of *Desulfovibrio* ("proto-*Desulfovibrio*") did not contain a hydrogenase. Such a conclusion has substantial implications for the metabolism of proto-*Desulfovibrio*. However, alternative possibilities again must be considered. In the absence of structural data, the possibility remains that hydrogenases employing different metal ions as cofactors might nevertheless be homologous (ignoring briefly the difficulty of continuous evolution from one to the other). The mechanistic divergence may then be adaptive, and therefore explained non-historically. Insufficient information is presently available to assess these possibilities.

There are some remarkable examples of non-homologous enzymes catalyzing seemingly identical reactions in a single organism. For example, the zinc-dependent alcohol dehydrogenases of *Saccharomyces cerevisiae* (1.1.1.1, in three isozymes, Adh I, II, and III) are apparently not homologous with an iron dependent alcohol dehydrogenase (Adh IV) found in the same organism [190, 191, 165]. This creates an evolutionary paradox. One would not expect an organism to invent de novo an enzyme catalyzing a reaction identical to that of an existing enzyme, even if it performed a different biological function. One would expect the existing enzyme to divergently evolve to occupy the new role (a process that apparently produced the three existing isozymes of the zinc-dependent enzyme). Naively, one might therefore argue that the second enzyme did not arise in yeast, but rather developed elsewhere and then transferred laterally to yeast. However, even this explanation is not entirely satisfactory, as an organism possessing two enzymes catalyzing identical reactions would be expected to lose one, via a deletion–replacement event, even if the two enzymes perform separate roles.

Thus, the presence of two mechanistically distinct enzymes suggests a separate functional role for each enzyme, one that depends explicitly on the choice of a metal cofactor. An obvious selectable difference for two enzymes employing different metal cofactors is that an organism containing both can survive in environments lacking one or the other metal. There is no independent evidence to support this hypothesis at present.

A General Pattern of Substrate Divergence Permits the Identification of "Families" of Proteins Catalyzing Similar Reactions on Structurally Different Substrates

Homology can also be detected between enzymes that catalyze reactions that are different in detail but chemically analogous. This suggests that substrate specificity can diverge within a series of enzymes that conserve mechanistic details. Proteases again provide textbook examples. For example, batroxobin (3.4.21.5) from the snake *Bothrops atrox*, *moojeni* has homology with serine proteases [192]. Aspartic proteases from a variety of organisms with different substrate specificities are generally homologous with pepsins. (3.4.4.1 [193–195]. Cathepsin L (3.4.22.15a) from chicken liver can be aligned with plant cysteine proteases, including papain, actinidin and aleurain [196].

Several other examples also show the ability of divergent evolution to create a family of proteins. One of the best examples, noted first by Rossmann and his colleagues and developed by Joernvall and his colleagues, are the alcohol dehydrogenases. Enzymes using ethanol, secondary alcohols, sorbitol, ribitol, and even glucose as substrates all appear to belong to an evolutionary "superfamily" of dehydrogenases sharing a dinucleotide binding domain, but having substantial differences in substrate specificity [197–199].

Aspartate transcarbamylases (ATCase, 2.1.3.2) from a variety of sources are homologous, as are ornithine transcarbamylases (OTCase, 2.1.3.3). However, these two classes of enzymes catalyzing analogous reactions are also homologous [200–206]. Fumarases, which catalyze the addition of water to fumarate, are homologous to aspartases, which catalyze the addition of ammonia to fumarate [207, 208]. Adenine phosphoribosyltransferase (APRT) from *E. coli* (2.4.2.7) is homologous with APRT from mouse, and also with xanthine-guanine PRT from man, xanthine-guanine PRT from *E. coli*, orotate PRT from *E. coli*, and glutamine PRT from *E. coli* [209]. Phosphatidylinositol synthase (2.7.8.11) from *Saccharomyces cerevisiae* is homologous with phosphatidylserine synthase, and phosphatidyl-1′-glycerol-3′-phosphate synthase [210]. Enzymes catalyzing the synthesis of cyclodextrins in *Bacillus* are strongly homologous with each other, and with human salivary amylase [211–213]. DNA polymerase (2.7.7.7) from mammals shows homology with terminal deoxynucleotidyltransferase [214].

Enzymes catalyzing different reactions that have common ancestors can be explained by assuming that one catalytic function predated the other. Here, one must assume that there existed in the past an organism that had either activity, but not both. Discussions then focus on which activity arose first. However, it is also possible that both activities were present in an ancestor in non-homologous forms, and one was deleted and replaced by cross-evolution of the other. Here, the deletion–replacement event is facilitated by the presence of an enzyme performing a chemically similar function that can be recruited following a few point mutations.

Comparative sequence analysis can, at least in principle, permit a distinction between these two explanations. If the most recent common ancestor had both enzyme A and B, then enzyme A in descendant 1 should be more similar to enzyme A of descendant 2 than to enzyme B of descendant 1. In contrast, if the homology of enzyme A and B in organism 1 reflects deletion–replacement events following the divergence of the progenote, then enzyme A from organism 1 should resemble enzyme B from organism 1 more than it does enzyme A from organism 2.

For example, it is clear that the most recent common ancestor of modern eukaryotes had both mitochondrial and cytoplasmic malate dehydrogenases. The mitochondrial enzymes are more closely related to mitochondrial enzymes from other organisms than to the cytoplasmic enzymes from the same organisms. This argument also illustrates the importance of an assumption that sequence convergence and lateral transfer are not frequent in the evolution of proteins [215].

However, Catalysts for Analogous Reactions have also Been Invented More than Once

Given the fact (vide supra) that some enzymes catalyzing identical reactions are not homologous, it is not surprising to find that not all enzymes catalyzing analogous reactions are homologous. The following examples serve as warning to those tempted to infer evolutionary homology from chemical analogy as the basis for model building [216].

For example, many pyridoxal-dependent enzymes acting on different substrates are detectably homologous. Biodegradative threonine dehydratase (4.2.1.16) from *E. coli* is homologous to the D-serine dehydratase from *E. coli* as well as to the biosynthetic threonine dehydratase [217, 218] despite the fact that these enzymes act on substrates with opposite absolute configurations. Likewise, cytoplasmic and mitochondrial aspartate aminotransferases (2.6.1.1) are homologous in eukaryotes and eubacteria (with at least 38% sequence identity) [219]; ornithine aminotransferase from rat appears to belong to this evolutionary family as well [220].

However, only local sequence similarities have been reported between tyrosine aminotransferase, ornithine aminotransferase, and mitochondrial aspartate aminotransferase, and these occur primarily around the lysine residue believed to bind pyridoxal [221]. Further, no significant similarities can be detected between aspartate amino transferases and the pyridoxal-dependent tyrosine 2-oxoglutarate aminotransferase (2.6.1.5) from rat [222]. In mitochondrial serine-pyruvate aminotransferase, it is difficult to identify any lysine residue that might be at the active site and surrounded by a correct consensus sequence that might indicate homology with other enzymes dependent on pyridoxal phosphate [223]. Thus, there may be more than one pedigree of enzymes dependent on pyridoxal cofactors.

Enzymes that employ thiamine for the decarboxylation of alpha keto acids are also homologous in many pathways. However, the 2-oxoglutarate dehydrogenase (1.2.4.2) subunit of the 2-oxoglutarate decarboxylase from *E. coli* shows very little sequence homology with the pyruvate dehydrogenase subunit of the pyruvate decarboxylase complex [224], even though the two share mechanism and apparently other protein subunits [225]. In contrast, the branched chain ketoacid dehydrogenase (1.2.4.4) appear to be homologous to pyruvate dehydrogenase [226, 227].

Similarly, some enzymes that catalyze the decarboxylation of amino acids use pyruvoyl residues generated in the polypeptide chain from serine residues, rather than pyridoxal cofactors. Two enzymes in this mechanistic class are adenosylmethionine decarboxylase (4.1.1.50) and histidine decarboxylase. Surprisingly, the first enzyme from *E. coli* displays no gross homology with the second from *Lactobacillus* [228].

Finally, methylases involved in eubacterial restriction–modification systems generally are homologous when acting on a particular nucleophilic center on DNA. However, there is little apparent homology between enzymes that alkylate different nucleophiles in DNA. For example, O^6-alkylguanine-DNA-alkyltransferase (2.1.1.72) from *E. coli* is homologous with a series of other DNA alkyl transferases to guanine [229]. Likewise, the Hhal methyltransferase (2.1.1.72) from *Haemophilus haemolyticus* recognizes GCGC and methylates on C, and is homologous with other prokaryotic enzymes methylating on the same site [230]. The Taql methylase (2.1.1.72) from *Thermus aquaticus* is homologous to other methyl-6-adenine methylases, the closest homologue is the PaeR7I methylase (where the recognition site is a subsequence of the Taql recognition site) [231]. However, there is no sequence homology between these groups of enzymes.

In a Few Cases, Broad Homology is seen even in Enzymes Catalyzing Steps in Secondary Metabolism

The distinction between primary and secondary metabolism could be useful if primary metabolic steps were more ancient than secondary metabolic steps. This may generally be the case. However, several examples are known where homology is observed between proteins from eukaryotes and eubacteria that catalyze steps in secondary metabolism. For example, the alpha amylases (3.2.1.1) from *Streptomyces limosus* show homology with mammalian and invertebrate alpha-amylases [232].

Such homologies are known to include arachaebacteria in a single case. The amino terminal sequence of the superoxide dismutase (1.15.1.1) from the archaebacterium *Halobacter cutirubrum* shows surprising homology with iron and manganese superoxide dismutases from eubacteria [233]. Assuming no lateral transfer, this implies that superoxide dismutase was present in the progenote, and suggests that superoxide, and presumably oxygen, was present in the environment of the progenote. This statement is important, as the geological date for the appearance of oxygen is approximately known from examination of sediments. If this conclusion is true, the progenote must have lived after this date.

In Some Cases, Homology is Possible Between Enzymes Catalyzing Quite Different Chemical Reactions

Sequence comparison are especially interesting when they detect homologies between proteins which share little obvious chemical or functional similarities. For example, a cytochrome P-450 reductase from pig seems homologous, at least in part, with glutathione reductase, ferredoxin reductase, and flavodoxin [234]. Lipoamide dehydrogenase, plasmid-encoded mercuric reductase, and human erythrocyte glutathione reductase are homologous [235]. Carbamoyl-phosphate synthetase may have arisen by the fusion of two genes, one for a glutaminase and one for a synthetase [236]. Human retinoic acid receptors and steroid receptors appear to be homologous to homoserine dehydrogenase from *E. coli* [237]. The small cytoplasmic 19S ribonucleoprotein from *Drosophila* is homologous to a multicatalytic proteinase from rat [238]. Human serum cholinesterase (3.1.1.8) has sequence similarity with bovine thyroglobulin [239]. Polynucleotide phosphorylase (2.7.7.8) from *E. coli* appears to be homologous to the domain of ribosomal protein S1 that binds to RNA [240].

In some cases, entire classes of proteins appear to be homologous. For example, many kinases and phosphotransferases appear to be derived from a common ancestor [241, 242]. The ATP binding site of adenylate kinase (2.7.4.3b) appears to be homologous to ras-encoded p21, F1-ATPase, and other nucleotide-binding proteins [243]. The K^+-ATPase of *Streptococcus faecalis* shows homology with the KdpB-protein of *Escherichia coli* [244].

In some cases, the homology can be rationalized in terms of a mechanism by which new enzymatic activities arise. For example, enzymes catalyzing different reactions in consecutive steps in a metabolic pathway are homologous. Keto-adipate enol-lactone hydrolase (3.1.1.24) and muconolactone cycloisomerase (5.3.3.4), enzymes catalyzing consecutive steps in the beta-ketoadipate pathway in *Pseudomonas putida*, appear to be homologous [245, 246]. Cystathionine-gamma-synthetase (4.2.99.9) and beta-cystathionase (4.4.1.8), enzymes catalyzing consecutive steps in the biosynthesis of methionine have been found to be homologous [247].

A common origin of enzymes catalyzing consecutive steps in biochemical pathways makes some sense, as a binding site for the product of the first enzyme might easily evolve to become a binding site for the substrate of the next. In its most general form, this argument implies that catalytic mechanism can drift more easily than binding-site specificity, an implication contradicting the discussion above. In the cases above, however, the consecutive reactions are chemically similar, making the proposal of homology chemically plausible.

In other cases, homologies suggest a correlation between structure and biological function. For example, poly(ADP-ribose) synthetase (2.4.2.30) from man has limited sequence homology with segments of transforming proteins such as c-fos and v-fos [248]. This suggests a physiological role for this protein in the control of cell division. Likewise, poly(ADP-ribose) synthetase (2.4.2.30b) from man contains sequences similar to those found in certain DNA binding proteins. This was interpreted as suggesting that polyribosylation is a response to DNA strand breaks during repair and replication [249].

A good example where homology has strongly indicated biological function centers on the homology between digestive ribonucleases and angiogenin, a blood growth factor that presumably assists tumor cell growth [250]. The sequence homology encouraged the search for (and eventual discovery of) ribonucleolytic activity in angiogenin (Fig. 11) [251]. This discovery in turn lends support to the hypothesis that RNA molecule, and extracellular RNA molecules in particular, are important messengers in eukaryotic development [252]. Evidence for such RNA molecules involved in tumor cell growth has been available for nearly 2 decades [253], and Wissler and his coworkers have recently characterized in depth an RNA-containing angiogenic factor [254]. Remarkably, angiogenin resembles turtle ribonuclease more closely than human ribonuclease in certain of its structural features (Fig. 11). This suggests that ribonucleases played a role in developmental biology well before they adopted the more familiar role as a digestive enzyme important in ruminants [255].

Searches for homology between enzymes catalyzing chemically non-analogous reactions is most productive when they are supported by a high quality data base. For example, sequence similarities suggest that acetyl-CoA carboxylase (6.4.1.2) from rat and carbamoyl phosphate synthetase (6.3.4.16) from yeast and rat are homologous [256]. The first enzyme catalyzes the ATP-dependent carboxylation of biotin, while the second catalyzes the ATP-dependent carboxylation of ammonia, reactions that are vaguely similar. However, a good data base points out that this is not the first proposal for an unusual homology involving carbamoyl phosphate synthetase. An earlier paper suggested that segments of mitochondrial aldehyde dehydrogenase (1.2.1.4) are homologous to carbamoyl phosphate synthetase as well [257].

One would not expect these three enzymes to be homologous a priori. Of the substrates for aldehyde dehydrogenase, only NAD^+ resembles a substrate for the other two enzymes; the adenine moiety resembles the ATP. Further, suggestions that NAD^+ binding domains are homologous to ATP binding domains can be found throughout the literature. Each example makes stronger the case that substantial divergence in chemical mechanism can be tolerated within a mechanistic class.

Manipulatable sequence data bases are valuable in other contexts. For example, Brenner recently noted that in some serine proteases, the AGY (where Y is a pyrimidine) codon was used for serine, while in others, the TCN (where N is any nucleoside base) codon was used [258]. One codon cannot be converted to the other without changing two bases at one time, a double mutation that Brenner presumed improbable. However, both codons could arise from the TGY codon (for cysteine). Thus, Brenner argued that the ancestral codon must have been a cysteine, and two branches of serine proteases must be descendants of a cysteine protease.

The argument has substantial intellectual appeal, and the conclusion it draws may in fact be correct. However, other examples of this phenomenon are known. For example, in the fumarase/aspartase alignment, a highly conserved sequence GlySerSerIleMetPro is absolutely conserved throughout an alignment with over 75% sequence divergence [205, 206], suggesting that the sequence is critically involved in the biological function of the protein. However, the serine codon is TGY in some genes and ACN in others; the divergence is seen within proteins that are otherwise 50% identical in sequence. This suggests that caution is appropriate in using the assumption that codon drift involving double mutation cannot occur rapidly as the basis for an argument for the structure of ancient proteins.

In this discussion, it is worth mentioning again that in many cases, sequence similarities used as the basis for a claim of homology are quite limited. In these cases, it is often difficult to tell whether the sequence similarities indicate homology, reflect convergent sequence evolution, or are simply accidental [259]. For example, lecithin-cholesterol acyltransferase (2.3.1.43) from man is not significantly similar with any other known sequence. However, it shares a Gly-Xxx-Ser-Xxx-Gly sequence with serine esterases, peptidases, and thioesterases [260]. This may reflect homology, convergence for a sequence that uniquely can specify a particular folded structure, or pure may accident. Likewise, the short polypeptide sequence between the redox reactive cysteines in ribonucleotide reductases dependent on B-12 and iron are quite similar (Cys-Glu-Gly-Gly-Ala-Cys in the first case, and Cys-Glu-Ser-Gly-Ala-Cys in the second) [261]. This similarity might indicate homology. However, it is clear that only a small number of sequences could accommodate the constraints imposed by function for this sequence, suggesting the strong possibility that the sequences of these two proteins have convergently evolved.

Indeed, searches have now been made to determine the frequency with which short repeating segments occur in random proteins [262]. In many cases where these repeating elements are found, similar secondary structure is observed. This implies that weak sequence similarities often reflect convergent sequence evolution, not homology, and care must be exercised accordingly.

When a statistical approach is used, correlations suggesting homology across a wide range of proteins can be found. For example, suggestions of homology has been found between the core subunits of RNA polymerase from *E. coli*, DNA primase, elongation factor Tu, F1ATPase, ribosomal protein L3, DNA polymerase I, T7 phage DNA polymerase, and MS2 phage RNA replicase beta subunit [263].

Finally, other types of data can suggest broad functional divergence in enzymes. For example, some enzymes contain cofactors that are unnecessary for the reaction that they catalyze. Hydroxynitrilase contains a flavin cofactor, even though it performs no redox reactions [264]. Glycogen phosphorylase contains pyridoxal phosphate, even though it catalyzes no pyridoxal-dependent chemistry [265]. Acetolactate synthetase contains a flavin, even though it catalyzes no redox reaction [266]. Phosphoribosylpyrophosphate amidotransferase (2.4.2.14) from *Saccharomyces cerevisiae* contains an iron–sulfur cluster essential for catalytic function; enzymes with both oxidized and reduced cofactor are catalytically active in avian, human, and *Bacillus* enzymes, suggesting no redox role for this cluster [267]. The presence of an unused cofactor suggests that the enzyme in question arose from an ancestral enzyme that used the cofactor, with substantial divergence in chemical role and mechanism. In this view, the unnecessary cofactor remains as a vestige of an ancient function for the enzyme.

5 Applications of Historical Models: Reconstructing the Breakthrough Organism

To illustrate the process by which structural and behavioral information can be used to construct models for very ancient forms of life, we consider four aspects of the metabolism of the breakthrough organism (Fig. 10).

5.1 The Breakthrough Organism Used DNA

DNA is the primary storage molecule for genetic information in the modern world. However, according to the model in Fig. 10, RNA was used for information storage in the first form of life. Did DNA emerge as the genetic material before or after the breakthrough? With the tools developed above, we can construct two competing models, one postulating that DNA emerged *after* the breakthrough, the other that DNA emerged *before* the breakthrough, and decide which model is preferable as a working hypothesis.

We consider the following facts, together with implications drawn from these facts:

a) Several DNA-dependent enzymes can be reliably assigned to the progenote. In particular, archaebacteria, eubacteria, and eukaryotes contain homologous DNA-dependent RNA polymerases [268]. This implies that the progenote had a DNA-dependent RNA polymerase, and therefore had DNA.
b) For organisms with DNA, endogenous ribonucleotide reductase (RNR), the enzyme that converts ribonucleotides into deoxyribonucleotides, appears to confer strong selective advantage. Trypanosomes and some viruses, organisms that are parasitic in most other respects, carry their own ribonucleotide reductase, rather than attempt to obtain deoxyribonucleotides in the diet [269]. Therefore, the presence of DNA implies the presence of a ribonucleotide reductase, while the absence of ribonucleotide reductase implies an absence of DNA. In particular, the conclusion (above) that the progenote contained DNA implies that the progenote contained a ribonucleotide reductase.
c) However, three (and possibly four) mechanistically different ribonucleotide reductases (RNR's) are known in the modern world [261, 270]. This implies that the RNR's in different kingdoms are not homologous, and makes it impossible to assign a chemical reaction mechanism to any RNR presumed to have been present in the progenote. Indeed, in the absence of evidence for point (a) and (b), it might be considered reasonable to suggest that the progenote did not contain *any* RNR, and that the mechanistically different RNR's in the modern world reflect the independent origin of several types of RNR following the divergence of the progenote.

Two alternative resolutions of this apparent contradiction can be considered. First, the progenotic RNR may have been a protein homologous to one of the modern RNR's. In this case, the two (or perhaps three) other types of RNR arose by protein-for-protein replacement events subsequent to the divergence of the progenote. Alternatively, the RNR in the progenote may have been a riboenzyme. In this case, contemporary RNR's all arose by protein-for-RNA replacement events [12].

An evaluation of the relative plausibility of these alternatives requires an evaluation of the relative facility of protein-for-protein deletion–replacement events compared to protein-for-RNA events. As discussed above, the frequency of deletion–replacement events depends on two factors: a) the extent to which the deletion is lethal, and b) the availability of genes for biological macromolecules that, after minor alteration, can assume the deleted function.

Ribonucleotide reductases are difficult to replace for both reasons. First, a deletion appears to be semi-lethal (point (a)). Second, even in modern metabolism, few proteins performing similar functions are available to replace a deleted RNR. Finally, and most significantly, there appears to be no selective advantage for replacing an enzyme with one mechanism by another. Thus, multiple protein-for-protein deletion–replacement events seems implausible for RNR. To the extent that a proteinaceous RNR was present in the progenote, we would expect it to have survived to the modern day in all lineages.

However, in view of the (presumed) advantage of protein over RNA as a catalyst, there should be a selective advantage for replacing an RNA ribonucleotide reductase by a protein. This makes more plausible a model for the progenote that includes a ribozymal ribonucleotide reductase. If we assume (above) that new catalytic RNA arose infrequently following the invention of translation, the riboenzyme in the progenote should be a descendant of a ribosomal RNA originating in the RNA world. This implies that the breakthrough organism contained a ribozymal RNR, and therefore contained DNA. Indeed, given the difficulty of deletion–replacement events for ribonucleotide reductase, it is not implausible that the riboenzyme form survived without replacement in the time separating the breakthrough organism and the progenote.

While this argument is based on parsimony and chemical structure, it is also plausible on general chemical and evolutionary grounds. As with any single metabolic step, ribonucleotide reduction requires less information than translation, and should have emerged before translation. Further, the chemical stability of DNA that makes it selectively advantageous for information storage in the modern world also should make it advantageous in the RNA world.

5.2 The Breakthrough Organism Biosynthesized Tetrapyrroles

We consider the following facts and their implications:

a) Two pathways exist for the biosynthesis of 5-aminolevulinate as the first step in the biosynthesis of tetrapyrroles (Fig. 15). One (the Shemin pathway) involves a chemically elegant condensation of succinyl-CoA and glycine

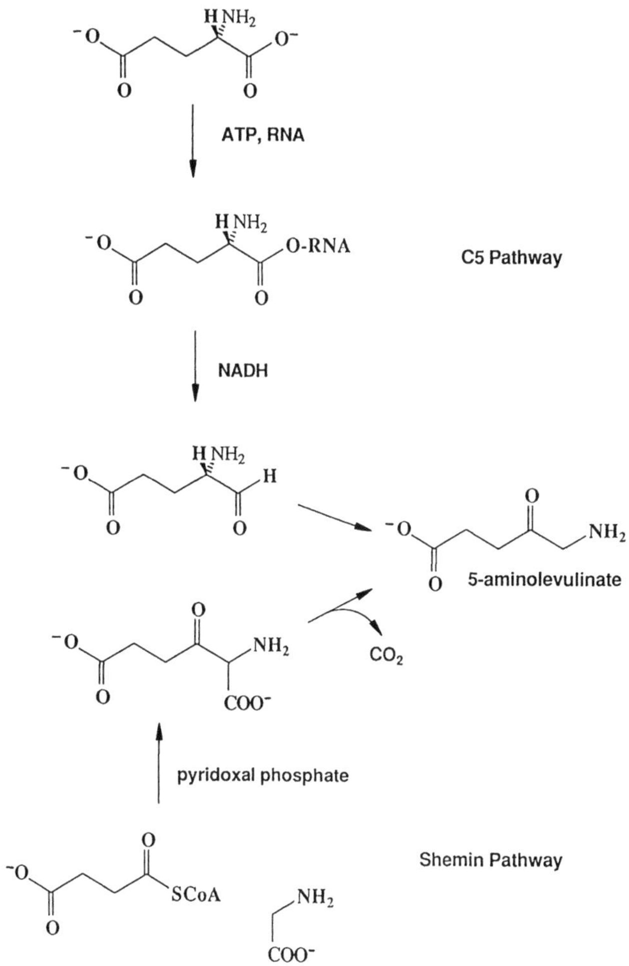

Fig. 15. Two pathways for the biosynthesis of 5-aminolevulinate. The C-5 pathway (top) begins with glutamate, and requires an RNA molecule as a cofactor. The Shemin pathway (bottom) starts with succinyl CoA and glycine

dependent on a pyridoxal cofactor. The second pathway (the C-5 pathway) involves the reduction of an ester of glutamic acid and RNA, followed by rearrangement of glutamate semialdehyde to give aminolevulinic acid [271].

b) Esters are not intrinsically advantageous intermediates in the biosynthesis of aldehydes from carboxylic acids; the reduction of anhydrides with phosphoric acid is chemically preferable, and is the route used in modern metabolism in virtually all cases. Even were esters desirable intermediates for such reactions, esters with RNA molecules seem to offer no intrinsic advantages over esters with simpler alcohols. Thus, if the C-5 pathway can be placed in the progenote, the involvement of RNA in the pathway strongly argues that the C-5 pathway (and its products) originated in the RNA world.

c) The C-5 pathway is used for the biosynthesis of chlorophyll in photosynthetic eukaryotes and in certain eubacteria [272], and has recently been described in archaebacteria for the biosynthesis of B-12 and Factor F430 [273]. Its presence in all three lineages descendant from the progenote allows an assignment of the C-5 path to the progenote.

d) The Shemin path is now known in two kingdoms descendant from the progenote, eukaryotes and eubacteria. Sequence evidence suggests that the enzymes from chicken, yeast, and *Bradyrhizobium japonicum* are homologous [274]. This contrasts with limited biochemical evidence suggesting that some pyridoxal-dependent 5-aminolevulinate synthetases from eukaryotes do not contain the active site lysine present in certain eubacterial enzymes dependent on a pyridoxal cofactor [275]. Further work is necessary to determine whether Shemin pathways in different kingdoms originated independently, were laterally transferred, or were present in the progenote. The pyridoxal enzymes should be homologous to other pyridoxal-dependent enzymes.

The presence in the progenote of an RNA molecule participating in the synthesis of 5-aminolevulinate, in a role not taking advantage of some unique chemical property of the RNA, supports an assignment of a riboenzyme synthesizing 5-aminolevulinate to the breakthrough organism. The role of this intermediate in the metabolism of the breakthrough organism remains uncertain, although a reasonable hypothesis is that it was used in the biosynthesis either of chlorophyll for photosynthesis, or B-12 for methanogenesis. Both roles are sufficiently central to metabolism of the respective organisms that deletion–replacement events are likely to be slow, explaining the conservation of the ribo-cofactor for tetrapyrrole biosynthesis in modern methanogens and photosynthetic organisms. There is at present no evidence that permits a strong hypothesis favoring one over the other, although weak hypotheses for both conclusions suggest contradicting outcomes to experiments involving the bioorganic details of the two pathways.

Whether or not oxygenic photosynthesis was developed in the RNA world is critical if we wish to assign geological time to the beginning and end of episodes depicted in Fig. 12. The origin of oxygenic photosynthesis can be dated at approximately 2.5 billion years B.P. If photosynthesis originated in the RNA world, the breakthrough must have occurred more recently than 2.5 billion years B.P. This implies that certain microfossils found in very old geological sediments are fossils of ribo-organisms.

5.3 The Breakthrough Organism Did Not Synthesize Fatty Acids

We consider the following facts and their implications:

a) Fatty acid synthase complexes from different organisms have different quaternary structures, stereospecificities, substrate specificities, and mechanism [276]. Archaebacteria do not seem to synthesize fatty acids at all [277]. Thus, as with ribonucleotide reduction, it is difficult to place fatty acid biosynthesis in the progenote.

b) However, unlike with ribonucleotide reductase, homologous enzymes are not yet known in the three kingdoms that use fatty acids as substrates. Therefore, there is no independent evidence that fatty acids were synthesized in the progenote.

c) The biosynthesis of fatty acids involves biotin and acyl carrier protein (ACP). Both almost certainly did *not* arise before translation. ACP is a product of translation; it must have emerged following the breakthrough. Chemical and structural features of biotin, reviewed a decade ago by Visser and Kellogg, strongly suggest that biotin emerged after protein catalysts [170].

These arguments suggest that fatty acid biosynthesis arose after the breakthrough. However, as lipids of some sort seem to be essential for living cells, an absence of fatty acid biosynthesis in the breakthrough organism creates a compelling need for an alternative source of lipids. This need can be satisfied by terpenes. Conversely, a successful assignment of terpenoid biosynthesis to the breakthrough organism removes a compelling need for fatty acid synthesis there.

5.4 The Breakthrough Organism Synthesized Terpenes

We consider the following facts and their implications:

a) Higher terpenes are biosynthesized via similar routes in all three kingdoms. This implies that the progenote biosynthesized terpenes, especially higher terpenes.

b) In the eubacterium *Rhodopseudomonas acidophila*, membranes contain terpenoids covalently joined to RNA fragments (Fig. 14), which serve as the polar part of the amphiphilic lipid molecule [278].

c) However, RNA is not uniquely suited as a polar group in this capacity. Indeed, in most lipids, polar components are not RNA. This suggests that RNA-conjugated lipids were present in the breakthrough organism.

The assignment of terpene biosynthesis to the progenote is stronger than the corresponding assignment of RNA-dependent tetrapyrrole biosynthesis. However, the involvement of vestigial RNA in terpene chemistry is less substantive than in tetrapyrrole biosynthesis; the hopanoid-RNA conjugate is presently known only in *Rhodopseudomonas*. Thus, it is more difficult to place terpene biosynthesis in the breakthrough organism. However, terpenoids are themselves constituents of chlorophyll. If chlorophyll can be placed in the breakthrough organism, this strengthens the assignment of terpene biosynthesis to the RNA world. Conversely, if terpene biosynthesis can be assigned to the breakthrough organism, this provides in ribometabolism an element necessary for the biosynthesis of chlorophyll.

6 Conclusion

This review is an effort to unify evolutionary and structural theories into a coherent model that addresses the behavior of biological macromolecules in a way that assists the bioorganic chemist designing experiments. Three points are clear from the discussion that deserve restatement in conclusion.

First, the effort is useful. An evolutionary perspective is valuable to those wishing to engineer proteins by site-directed mutagenesis, provides the best approach presently available for predicting the tertiary structure of proteins from sequence data, and is central to any effort to organize the volume of sequence data now emerging.

However, to be useful, the model building must be as logically rigorous as available data allow. It is no longer either necessary or acceptable to discuss evolution, enzymatic behavior, or natural selection at the macromolecular level in imprecise terms or ideas. Models that are logically formal models are now possible [12–16], and are essential if they are to guide experimental research.

Third, well-constructed models are tentative; only a few contradicting experimental data can force their revision. The experiments needed to obtain these data are clear from the model themselves, making the model a powerful engine to drive new experimental effort.

7 References

1. Kornberg A (1987) Biochem. 26: 6888
2. von Heijne G (1988) Nature 333: 605
3. Faulkner DV, Jurka J (1988) Trends Biochem. Sci. 13 : 321
4. Chothia C, Lesk AM (1986) EMBO J. 5: 823
5. Blundell TL, Sibanda BL, Sternberg MJE, Thornton JM (1987) Nature 326: 347
6. Knowles JR (1987) Science 236: 1252
7. Ulmer KM (1983) Science 219: 666
8. Fersht AR, Leatherbarrow RJ, Wells TC (1987) Biochem. 26: 6030
9. Jencks WP (1987) Cold Spring Harbor Symp. Quant. Biol. 52: 65
10. Jencks (1975) Adv. Enzymol. 43: 219
11. Benner SA (1988) Chimia 42: 309
12. Benner SA, Ellington AD (1988) CRC Crit. Rev. Biochem. 23: 369
13. Benner SA, Glasfeld A, Piccirilli JA (1989) In: Eliel EL, Wilen SH (eds) Topics in stereochemistry 19, Wiley, New York, p. 127
14. Benner SA (1989) Chem. Rev. 89: 789
15. Benner (1987) Cold Spring Harbor Symp. Quant. Biol. 52: 53
16. Benner SA (1988) In: Benner SA (ed) Redesigning the molecules of life, Springer, Berlin Heidelberg New York, p 115
17. Dequard-Chablat M (1986) J. Biol. Chem. 261: 4117
18. Pietruszko R (1982) Meth. Enzymol. 89: 428
19. Stalker DM, Hiatt WR, Comai L (1985) J. Biol. Chem. 260: 4724
20. Both GW, Shi CH, Kilbourne ED (1983) Proc. Nat. Acad. Sci. 80: 6996
21. Wells JA, Estell DA (1988) Trends Biochem. Sci. 13: 291–297
22. Miller JH (1979) J. Mol. Biol. 131: 249
23. Shortle D, Lin B (1985) Genetics 110: 539
24. Serpersu EH, Shortle D, Mildvan AS (1986) Biochem. 25: 68
25. Cone JL, Cusumano CL, Taniuchi H, Anfinsen C (1971) J. Biol. Chem. 246: 3103

26. Hermes JD, Blacklow SC, Knowles JR (1987) Cold Spring Harbor Symp. Quant. Biol. 52: 597
27. Strauss D, Raines R, Kawashima E, Knowles JR, Gilbert W (1985) Proc. Nat. Acad. Sci. 82: 2272
28. Beintema JJ, Fitch WM, Carsana A (1986) Mol. Biol. Evol. 3: 262
29. Capasso, S, Giordano F, Mattia CA, Mazzarella L, Zagari A (1983) Biopolymers 22: 327
30. Piccoli R, D'Alessio G (1984) J. Biol. Chem. 259: 693
31. Vescia S, Tramontano D (1981) Mol. Cell. Biochem. 36: 125
32. Levy CC, Karpetsky TP (1981) In: Holcenberg JS, Roberts J (eds) Enzymes as drugs, Wiley, New York, p 103
33. Palmer KA, Scheraga HA, Riordan JF, Vallee BL (1986) Proc. Nat. Acad. Sci. 83: 1965
34. Shapiro R, Riordan JF, Vallee BL (1986) Biochem. 25: 3527
35. Wills C, Joernvall H (1979) Eur. J. Biochem. 99: 323
36. Branden CL, Joernvall H, Eklund H, Furugren B (1975) The Enzymes, 3rd ed 11: 182
37. Joernvall H, von Bahr-Lindstroem H, Jany K-D, Ulmer W, Froeschle M (1984) FEBS Lett. 165: 190
38. Allemann RK, Hung R, Benner SA (1988) J. Am. Chem. Soc. 110: 5555
39. Craik CS, Largman C, Fletcher T, Roczniak S, Barr PJ, Fletterick R, Rutter WJ (1985) Science 228: 291
40. Kreitman M (1983) Nature 304: 412
41. Jacq C, Miller JR, Brownlee GG (1977) Cell 12: 109
42. Vanin EF (1984) Biochem. Biophys. Acta 782: 231
43. Miyata T, Hayashida H (1981) Proc. Nat. Acad. Sci. 78: 5739
44. Li W-H, Wu, Chung-I L Chi-Cheng (1984) J. Molec. Evol. 21: 58
45. Li W-H, Gojobori T, Nei M (1981) Nature 292: 237
46. Gitlin D, Gitlin JD (1975) In: Putman FW (ed) The plasma proteins, Academic, New York
47. Perler F, Efstradiatis A, Lomedico P, Gilbert W, Koler R, Dodgson J (1980) Cell 20: 556
48. Crabtree GR, Comeau CM, Fowlkes DM, Fornace Jr AJ, Malley JD, Kant JA (1985) J. Mol. Biol. 185: 1
49. Minghetti PP, Law SW, Dugaiczyk A (1985) Mol. Biol. Evol. 2: 347
50. Hendriks W, Leunissen J, Nevo E, Bloemendal H, de Jong WW (1987) Proc. Nat. Acad. Sci. 84: 5320
51. Lewontin RC (1985) Ann. Rev. Genet. 19: 81
52. Weatherall DJ, Clegg JB (1976) Ann. Rev. Genet. 10: 157
53. Allison AC (1955) Cold Spring Harbor Symp. Quant. Biol. 20: 239
54. Yang Y-C, Ciarletta AB, Temple PA, Chung MP, Kovacic S, Witek-Giannotti JS, Leary AC, Kriz R, Donahue RE, Wong GG, Clark SC (1986) Cell 47: 3
55. Hill RE, Hastie ND (1987) Nature 326: 96
56. Stewart C-B, Schilling JW, Wilson AC (1987) Nature 330: 401
57. Dayoff MO (1972) Atlas of protein sequence and structure, Silver Spring, National Biomedical Research Foundation
58. Lewin R (1986) Science 232: 578
59. Diamond JM (1986) Nature 321: 565
60. Dykhuizen DE (1978) Evolution 32: 125
61. Koch AL (1983) J. Mol. Evol. 19: 455
62. Li W-H (1984) Mol. Biol. Evol. 1: 213
63. Benner SA (1989) In: Liebman JF, Greenberg A (eds) Molecular structure and energetics, Vol 9, VCH, Deerfield Beach, FL 27–74
64. Domingo E, Sabo D, Taniguchi T, Weissmann C (1978) Cell 13: 735
65. Sharp PM, Rogers MS, McConnell DJ (1985) J. Mol. Evol. 21: 150
66. Grantham R, Gautier C, Gouy M, Jocobzone M, Mercier R (1981) Nucl. Acids. Res. 9: r43
67. Gouy M, Gautier C (1982) Nucl. Acids. Res. 10: 7056
68. Konigsberg W, Godson GN (1983) Proc. Nat. Acad. Sci. 80: 687
69. Bennetzen J, Hall B (1982) J. Biol. Chem. 257: 3026
70. Grosjean H, Fiers W (1982) Gene 18: 199
71. Blaisdell BE (1983) J. Mol. Evol. 19: 226
72. Rubin CM, Houck CM, Deininger PL, Friedmann T, Schmid CW (1980) Nature 284: 372
73. Modiano G, Battistuzzi G, Motulsky AG (1981) Proc. Nat. Acad. Sci. 78: 1110
74. Lipman DJ, Wilbur WJ (1983) J. Mol. Biol. 163: 363
75. Lanave C, Preparata G, Saccone C (1985) J. Mol. Evol. 21: 346
76. Ahern TJ, Casal JI, Petsko GA, Klibanov AM (1987) Proc. Nat. Acad. Sci. 84: 675

77. Ellington AD, Benner SA (1987) J. Theor. Biol. 127: 491
78. Ellington AD (1988) Thesis, Harvard University
79. Place AR, Powers DA (1979) Proc. Nat. Acad. Sci. 76: 2354
80. Borgman V, Moon TW (1975) Can. J. Biochem. 53: 998
81. Hennessey Jr JP, Siebenaller JF (1987) Biochem. Biophys. Acta. 982: 285
83. Koehn RK (1969) Science 163: 943
84. Merritt RB (1972) Am. Nat. 106: 173
85. Day TH, Hiller PC, Clarke B (1974) Biochem. Genet. 11: 141
86. Day TH, Needham L (1974) Biochem. Genet. 11: 167
87. Glasfeld A (1988) Thesis, Harvard University
88. Zimmermann T, Kulla HG, Leisinger T (1983) Experientia 39: 1429
89. Campbell HH, Lengyel JA, Langridge J (1973) Proc. Nat. Acad. Sci. 70: 1841
90. Boronat A, Aguilar J (1981) Biochim. Biophys. Acta 672: 98
91. Burleigh BD Jr, Rigby PWJ, Hartley BS (1974) Biochem. J. 143: 341
92. Betz JL, Brown PR, Smyth MJ, Clarke PH (1974) Natue 247: 261
93. Hall A, Knowles JR (1976) Nature 264: 803
94. Wills C (1976) Nature 261: 26
95. Brooker RJ, Wilson TH (1985) Proc. Nat. Acad. Sci. 82: 3959
96. Murray M, Osborne S, Sinnott ML (1983) J. Chem. Soc. Perkin. Trans. II 1595
97. Hall BG, Yokoyama S, Calhoun DH (1984) Mol. Biol. Evol. 1: 109
98. Llewellyn DJ, Daday A, Smith GD (1983) J. Biol. Chem. 255: 2077
99. Matsumura M, Yasumura S, Aiba S (1986) Nature 323: 356
100. Branden C-I, Schneider G, Lindqvist Y, Andersson I, Knight S, Lorimer G (1987) Cold Spring Harbor Sym. Quant. Biol. 52: 491
101. Hartman FC, Soper TS, Niyogi SK, Mural RJ, Foote RS, Mitra S, Lee EH, Machanoff R, Larimer FW (1987) J. Biol. Chem. 262: 3496
102. Barber J (1987) Natute 325: 663
103. Miziorko HM, Lorimer GH (1983) Ann. Rev. Biochem. 52: 507
104. Andrews TJ, Lorimer GH (1978) FEBS Lett. 90: 1
105. Oppenheimer NJ, Marschner TM, Malver O, Kam B (1986) Enzyme reaction mechanism: Stereochemistry Elsevier, New York, p 15
106. Benner SA (1982) Experientia 38: 633
107. Benner SA, Stackhouse J (1982) In: Green BS, Ashani Y, Chipman D (eds) Chemical approaches to understanding enzyme catalysis, Elsevier, New York, p 32
108. Nambiar KP, Stauffer DM, Kolodziej PA, Benner SA (1983) J. Am. Chem. Soc. 105: 5886
109. Benner SA, Nambiar KP, Chambers GK (1985) J. Am. Chem. Soc. 107: 5513
110. Baker TA, Grossman AD, Gross CA (1984) Proc. Nat. Acad. Sci. 81: 6779
111. Goff SA, Casson LP, Goldberg AL (1984) Proc. Nat. Acad. Sci. 81: 6647
112. Zale SE, Klibanov AM (1983) Biotech. Bioeng. 25: 2221
113. Zuber H (1978) In: Friedman SM (ed) Biochemistry of thermophily, Academic, New York, p 267
114. Ahern TJ, Klibanov AM (1985) Science 228: 1280
115. Hamilton PT, Reeve JN (1985) J. Mol. Evol. 22: 351
116. Zuber H (1988) Biophys. Chem. 29: 171
117. Matthews BW, Nicholson H, Becktel WJ (1987) Proc. Nat. Acad. Sci. 84: 6663
118. Matthews BW (1987) Biochem. 26: 6885
119. Benner SA (1989) Adv. Enzym. Reg. 28: 219
120. Fasman GD (ed) (in press) Prediction of protein structure and the principles of protein conformation, Plenum, NY
121. Eisenberg D (1985) Ann. Rev. Biochem. 53: 595
122. Gould SJ (1980) The Panda's Thumb W W Norton, NY
123. Quack M (1986) Chem Phys. Lett. 132 147
124. Lake JA (1987) Cold Spring Harbor Symp. Quant. Biol. 52: 839
125. Fitch W (1977) Am. Nat. 111: 223
126. Kisakuerek MV, Leeuwenberg AJM, Hesse M (1983) Alkaloids: Chemical and Biological Perspectives 1: 211
127. Beintema JJ (1987) Life Chem. Reports 4: 333
128. Olsen GJ (1987) Cold Spring Harbor Symp. Quant. Biol. 52: 825
129. Lake JA (1987) Mol. Biol. Evol. 4: 167
130. Ebbole DJ, Zalkin H (1987) J. Biol. Chem. 262: 8274

131. Reeve JN, Beckler GW, Brown JW, Cram DS, Haas ES, Hamilton PI, Morris CJ, Sherf BA, Weil CP (1987) in Microbial Growth Cl Compounds, Proc 5th Int Symp, van Verseveld, Duine, JA, 255–260
132. Takeishi K, Kaneda S, Ayusawa D, Shimizu K, Gotoh O, Seno T (1985) Nucl. Acids. Res. 13: 2035
133. Phillips GJ, Kushner SR (1987) J. Biol. Chem. 262: 455
134. Doonan S, Barra D, Bossa F (1984) Int. J. Biochem. 16: 1193
135. Schuster W, Brennicke A (1987) EMBO J. 6: 2857
136. Timmis JN, Scott NS (1984) Trends Biochem. Sci. 9: 271
137. Reanney DC, Roberts WP, Kelly WJ (1982) In: Bull AT, Slater JH (eds) Microbial interactions and communities, vol 1, Academic, New York, p 287
138. Sogin ML, Ingold A, Karlok M, Nielsen H, Engberg (1986) EMBO J. 5: 3625
139. Michel F, Cummings DJ (1985) Curr. Genet. 10: 69
140. Lang BF (1984) EMBO J. 3: 2129
141. Gilbert W, Marchionni M, McKnight G (1986) Cell 46: 151
142. Marchionni M, Gilbert W (1986) Cell 46: 133
143. Duester G, Smith W, Bilanchone V, Hatfield GW (1986) J. Biol. Chem. 261: 2027
144. Dennis ES, Gerlach WL, Pryor AJ, Bennetzen JL, Inglis A, Llewellyn D, Sachs MM, Ferl RL, Peacock WJ (1984) Nucl. Acids Res. 12: 3983
145. Mounier N, Prudhomme JC (1986) Biochimie 68: 1053
146. Crabtree GR, Comeau C, Fowlkes DM, Fornace AJ, Malley JD, Kant JA (1985) J. Mol. Biol. 85: 1
147. Higashi Y, Yoshida H, Yamane M, Gotoh O, Yoshiaki FJ (1986) Proc. Nat. Acad. Sci. 83: 2841
148. Govind S, Bell PA, Kemper B (1986) DNA 5: 371
149. Lowe DG, Capon DJ, Delwart E, Sakaguchi AY, Naylor SL, Goeddel DV (1987) Cell 48: 137
150. Busslinger M, Rusconi S, Birnstiel ML (1982) EMBO J. 1: 27
151. King A, Melton DA (1987) Nucl. Acids Res. 15: 10469
152. Carlson TA, Chelm BK (1986) Nature 322: 568
153. Brownlee AG (1986) Nature 323: 585
154. Heinemeyer W, Buchmann I, Tone DW, Windass JD, Alt-Moerbe J, Weiler EW, Botz T, Schroeder J (1987) Mol. Gen. Genet. 210: 156
155. Bannister JV, Parker MW (1985) Proc. Nat. Acad. Sci. 82: 149
156. Lewin R (1985) Science 227: 1020
157. Leunissen JAM, de Jong WW (1986) J. Mol. Evol. 23: 250
158. Richardson J (1981) Adv. Prot. Chem. 34: 167
159. Chothia C (1988) Nature 333: 598
160. Benner SA, Ellington AD (1988) Nature 329: 295
161. Benner SA, Allemann RK, Ellington AD, Ge L, Glasfeld A, Leanz GF, Krauch T, MacPherson LJ, Moroney S, Piccirilli JA, Weinhold E (1987) Cold Spring Harbor Sym. Quant. Biol. 52: 53
162. Gilbert W (1986) Nature 319: 818
163. Woese CR (1987) Microbiol. Rev. 51: 221
164. Woese CR (1967) The origins of the genetic code, Harper and Row, New York
165. Crick F (1968) J. Mol. Biol. 38: 367
166. Orgel LE (1968) J. Mol. Biol. 38: 381
167. Lehninger A (1972) Biochemistry, Worth, New York
168. Usher DA, McHale AH (1976) Proc. Nat. Acad. Sci. 73: 1149
169. White III HB (1976) J. Mol. Evol. 7: 101
170. Visser CM, Kellogg RM (1978) J. Mol. Evol. 11: 171
171. Bass BL, Cech TR (1984) Nature 308: 820
172. Guerrier-Takada C, Altman S (1983) Science 223: 285
173. Darnell JE, Doolittle WF (1986) Proc. Natl. Acad. Sci. 83: 1271
174. Weiner AM, Maizels N (1987) Proc. Nat. Acad. Sci. 84: 7383
175. Alberts BM (1986) Amer. Zool. 26: 781
176. Cantoni GL (1962) Methods Enzymol. 5: 743
177. Siu PML, Wood HG (1962) J. Biol. Chem. 237: 3044
178. Maizels N, Weiner AM (1987) Nature 330: 616
179. Sedgwick S, Johnston L (1987) Nature 329: 109
180. Ciechanover A, Wolin SL, Steitz JA, Lodish HF (1985) Proc. Nat. Acad. Sci. 82: 1341
181. Ellington AD, Benner SA, Tauer A (1989) Proc. Nat. Acad. Sci. 86 : 7054
182. Eschenmoser A (1988) Angew. Chem. 27: 5

183. Miller SL (1955) J. Am. Chem. Soc. 77: 2351
184. Itoh T, Kumazaki T, Sugiyama M, Otaka E (1988) Biochem. Biophys. Acta 949: 110
185. Mauguen Y, Hartley RW, Dodson EJ, Dodson GG, Borcogne G, Chothia C, Jack A (1982) Nature 297: 162
186. Hirata H, Fukazawa T, Negoro S, Okada H (1986) J. Bact. 166: 722
187. Matsuda A, Toma K, Komatsu K (1987) J. Bact. 169: 5821
188. Matthews DA, Smith SL, Baccanari DP, Burchall JJ, Oatley SM, Kraut J (1986) Biochem. 25: 4194
189. Menon NK, Peck HD, Le Gall J, Przybyla AE (1987) J. Bact. 169: 5401
190. Williamson VM, Paquin CE (1987) Mol. Gen. Genet. 209: 374
191. Conway T, Sewell GW, Osman YA, Ingram LO (1987) J. Bact. 169: 2591
192. Itoh N, Tanaka N, Mihashi S, Yamashina I (1987) J. Biol. Chem. 262: 3132
193. Takahashi K (1987) J. Biol. Chem. 262: 1468
194. Delaney R, Wong RNS, Meng G, Wu N, Tang J (1987) J. Biol. Chem. 262: 1461
195. Barkholt V (1987) Eur. J. Biochem. 167: 327
196. Ishidoh K, Towatari T, Imajoh S, Kawasaki H, Kominami E, Katunuma N, Suzuki K (1987) FEBS Lett. 223: 69
197. Rossmann MJ, Liljas A, Branden CI, Banaszak LJ (1975) The Enzymes 11: 61
198. Joernvall H, von Bahr-Lindstroem H, Jany KD, Ulmer W, Froeschle M (1984) FEBS Lett. 165: 190
199. Eklund H, Horjales E, Joernvall H, Branden CI, Jeffery J (1985) Biochem. 24: 8005
200. Michaels G, Kelln RA, Nargang FE (1987) Eur. J. Biochem. 166: 55
201. Shigesada K, Stark GR, Maley JA, Niswander LA, Davidson JN (1985) Mol. Cell. Biol. 5: 1735
202. Lerner CG, Switzer RL (1986) J. Biol. Chem. 261: 11156
203. Upshall A, Gilbert T, Saari G, O'Hara PJ, Weglenski P, Berse B, Miller K, Timberlake WE (1986) Mol. Gen. Genet. 204: 349
204. Baur H, Stalon V, Falmagne P, Luethi E, Haas D (1987) Eur. J. Biochem. 166:111
205. Huygen R, Crabeel M, Glansdorff N (1987) Eur. J. Biochem. 166: 371
206. Takiguchi M, Murakami T, Miura S, Mori M (1987) Proc. Nat. Acad. Sci. 84: 6136
207. Kinsella BT, Doonan S (1986) Bioscience Rep. 6: 921
208. Woods SA, Miles JS, Roberts RE, Guest JR (1986) Biochem. J. 237: 547
209. Hershey HV, Taylor MW (1986) Gene 43: 287
210. Nikawa J, Kodaki T, Yamashita S (1987) J. Biol. Chem. 262: 4876
211. Hamamoto T, Kaneko T, Horikoshi K (1987) Agric. Biol. Chem. 51: 2019
212. Takano T, Fukuda M, Monma M, Kobayashi S, Kainuma K, Yamane K (1986) J. Bact. 166: 1118
213. Kimura K, Kataoka S, Ishii Y, Takano T, Yamane K (1987) J. Bact. 169: 4399
214. Matsukage A, Nishikawa K, Ooi T, Seto Y, Yamaguchi M (1987) J. Biol. Chem. 262: 8960
215. Birktoft JJ, Fernley RT, Bradshaw RA, Banaszak LJ (1982) Proc. Nat. Acad. Sci. 79: 6166
216. Chari RVJ, Whitman CP, Kozarich JW, Ngai KL, Ornston LN (1987) J. Am. Chem. Soc. 109: 5514
217. Datta P, Goss TJ, Omnaas JR, Patil RV (1987) Proc. Nat. Acad. Sci. 94: 393
218. Parsot C (1986) EMBO J. 5: 3013
219. Kondo K, Wakabayashi S, Kagamiyama H (1987) J. Biol. Chem. 262: 8648
220. Inana G, Totsuka S, Redmond M, Dougherty T, Nagle J, Shiono T, Ohura T, Kominami E, Katunuma N (1986) Proc. Nat. Acad. Sci. 83: 1203
221. Mueckler MM, Pitot HC (1985) J. Biol. Chem. 260: 12993
222. Grange T, Guenet C, Dietrich JB, Chasserot S, Fromont M, Befort N, Jami J, Beck G, Pictet R (1985) J. Mol. Biol. 184: 347
223. Oda T, Miyajima H, Suzuki Y, Ichiyama A (1987) Eur. J. Biochem. 168: 537
224. Darlison MG, Spencer ME, Guest JR (1984) J. Biochem. 141: 351
225. Steginsky CA, Gruys KJ, Frey PA (1985) J. Biol. Chem. 260: 13690
226. Zhang B, Kuntz MJ, Goodwin GW, Harris RA, Crabb DW (1987) J. Biol. Chem. 262: 15220
227. Dahl H-HM, Hunt SM, Hutchison WM, Brown GK (1987) J. Biol. Chem. 262: 7398
228. Tabor CW, Tabor H (1987) J. Biol. Chem. 262: 16037
229. Potter PM, Wilkinson MC, Fitton J, Carr FJ, Brennand J, Cooper DP, Margison GP (1987) Nucl. Acids Res. 15: 9177
230. Caserta M, Zacharias W, Nwankwo D, Wilson GG, Wells RD (1987) J. Biol. Chem. 262: 4770

231. Slatko BE, Benner JS, Jager-Quinton T, Moran LS, Simcox TG, van Cott EM, Wilson GG (1987) Nucl. Acids Res. 15: 781
232. Long CM, Virolle MJ, Chang SY, Chang S, Bibb MJ (1987) J. Bact. 169: 5745
233. May BP, Dennis PP (1987) J. Bact. 169: 1417
234. Haniu M, Izanagi T, Miller P, Lee TD, Shively JE (1986) Biochem. 25: 7906
235. Stephens PE, Lewis HM, Darlison MG, Guest JR (1983) Eur. J. Biochem. 135: 519
236. Nyunoya H, Broglie KE, Lusty CJ (1985) Proc. Nat. Acad. Sci. 82: 2244
237. Baker ME (1988) Biochem. J. 255: 748
238. Falkenburg PE, Haass C, Kloetzel PM, Niedel B, Kopp F, Kuehn L, Dahlmann B (1988) Nature 331: 190
239. Lockridge O, Bartels CF, Vaughan TA, Wong CK, Norton SE, Johnson LL (1987) J. Biol. Chem. 262: 549
240. Regnier P, Grunberg-Manago M, Portier C (1987) J. Biol. Chem. 262: 63
241. Levin DE, Hammond CI, Ralston RO, Bishop JM (1987) Proc. Nat. Acad. Sci. 84: 6035
242. Brenner S (1987) Nature 329: 21
243. Fry DC, Kuby SA, Mildvan AS (1986) Proc. Nat. Sci 83: 907
244. Solioz M, Mathews S, Fuerst P (1987) J. Biol. Chem. 262: 7358
245. Yeh WK, Davis G, Fletcher P, Ornston LN (1978) J. Biol. Chem. 253: 4920
246. Yeh W-K, Fletcher P, Ornston LN (1980) J. Biol. Chem. 255: 6342
247. Belfaiza J, Parsot Cl, Martel A, Bouthier de la Tour C, Margarita D, Cohen GN, Saint-Girons I (1986) Proc. Nat. Acad. Sci. 83: 867
248. Kurosaki T, Ushiro H, Mitsuuchi Y, Suzuki S, Matsuda M, Matsuda Y, Katunuma N, Kangawa K, Matsuo H, Hirose T, Inayama S, Shizuta Y (1987) J. Biol. Chem. 262: 15990
249. Cherney BW, McBride OW, Chen D, Alkhatib H, Bhatia K, Hensley P, Smulson ME (1987) Proc. Nat. Acad. Sci. 84: 8370
250. Strydom DJ, Fett JW, Lobb RR, Alderman EM, Bethune JL, Riordan JF, Vallee BL (1985) Biochem. 24: 5494
251. Shapiro R, Riordan JF, Vallee BL (1986) Biochem. 25: 3527
252. Benner SA (1988) FEBS Lett. 233: 225
253. Folkman J, Merler E, Abernathy C, Williams G (1971) J. Exp. Med. 133: 275
254. Wissler JH, Logemann E, Meyer HE, Kruetzfeldt B, Hoeckel M, Heilmeier Jr. LMG (1986) Protides Biol Fluids 34: 525
255. Barnard EA (1969) Nature 221: 340
256. Lopez-Casillas F, Bai DH, Luo, X, Kong IS, Hermodson MA, Kim KH (1988) Proc. Nat. Acad. Sci. 85: 5784
257. Hempel J, Hoeoeg J-O, Jornvall H (1987) FEBS Lett. 222: 95
258. Brenner S (1988) Nature 334: 528
259. Doolittle RF (1981) Science 214: 149
260. Yang C, Manoogian D, Pao Q, Lee F, Knapp RD, Gotto AM Jr, Pownall HJ (1987) J. Biol. Chem. 262: 3086
261. Lin ANI, Ashley GW, Stubbe J (1987) Biochem. 26: 6905
262. Wuilmart C, Delhaise P, Urbain J (1982) Bio. Systems 15: 221
263. Ohnishi K (1985) 13th Symposium Nucleic Acids Chem, pp 253–256
264. Jorns MS (1985) Biochem. Biophys. Acta 830: 30
265. Butler PE, Cookson EJ, Beynon RJ (1985) Biochem. Biophys. Acta. 847: 316
266. LaRossa RA, Falco SC, Mazur BJ, Livak KJ, Schloss JV, Smulski DR, Van Dyk TK, Yadav NS (1987) ACS Symp Ser. 334: 190
267. Mantsala P, Zalkin H (1984) J. Biol. Chem. 260: 8478
268. Gropp F, Reiter WD, Sentenac A, Zillig W, Schnabel R, Thomm M, Stetter KO (1986) System Appl. Microbiol. 7: 9518
269. Ator MA, Stubbe J, Spector T (1986) J. Biol. Chem. 261: 3595
270. Hogenkamp HPC, Follmann H, Thauer RK (1987) FEBS Lett. 219: 197
271. Huang DD, Wang WY, Gough SP, Kannangara CG (1984) Science 225:1482
272. Kannangara CG, Gough SP, Bruyant P, Hoober JK, Kahn A, von Wettstein D (1988) Trends Biochem. Sci. 13: 139
273. Gilles H, Jaenchen R, Thauer RK (1983) Arch. Microbiol. 135: 237
274. McClung CR, Somerville JE, Guerinot ML, Chelm BK (1987) Gene 54: 13
275. Nandi DL (1978) Arch. Biochem. Biophys. 188: 266

276. McCarthy AD, Hardie DG (1984) Trends Biochem. Sci. 9: 60
277. Ross HNM, Collins MD, Tindall BJ, Grant WD (1981) J. Gen. Microbiol. 123: 75
278. Neunlist S, Rohmer M (1985) Biochem. J. 228: 769
279. Crawford IP, Niermann T, Kirschner K (1987) Proteins 2: 219
280. Rich A (1962) Horizons in Biochem. Kasha M, Pullman Beds. NY Academic Press, 103
281. Schram G, Grötsch H, Pollmann W (1961) Angew Chem 73: 619

Biomimetic Ion Transport with Synthetic Transporters

Thomas M. Fyles
Department of Chemistry, University of Victoria, Victoria, B.C., Canada V8W 2Y2

Membrane structure, and transport function play a central role in biochemical systems. Despite the complexity of natural systems, some of the structural and functional features of transporters can be imitated by simple chemical models in artificial systems. Such mimics serve to illustrate the basic constraints on transporters, artificial or natural, and serve to define structural and energetic essentials for transport. Mimics can act in two roles: mimics of structure or mimics of function. Carrier molecules mimic the structures of ionophore antibiotics, but can also illustrate gradient pumping, active transport and regulatory functions of natural systems. Polymer membrane systems provide insight into relay type transporters and illustrate higher order regulatory effects. Bilayer membranes provide the most realistic systems for biomimetic ion transport. Recent progress in the synthesis of pores and unimolecular ion channels, reveals that simple models give highly active transporters. Biomimetic ion transport studies also provide new systems for applications in separations and sensor developments.

Dedicated to the memory of a pioneer of biomimetic chemistry, Professor Iwao Tabushi

1 Introduction

Biological membranes provide a vast array of functions for a cell [1]. Foremost is the role of the membrane as the cell boundary which separates and protects the internal biochemical apparatus of the cell from the environment. The plasma membrane serves as a general barrier and regulates the import of nutrients and the export of cell products. Regulation and transport is controlled by specific transport proteins which traverse the membrane. In eukaryotes there is further subcellular differentiation by different types of membranes: mitochondria, endoplasmic reticulum, chloroplasts in plants. Enzymes and enzyme products in isolated compartments are protected from one another yet are integrated into the functioning of the cell via regulation of transport between compartments. Many biochemical functions are achieved by membrane bound enzymes and higher order processes such as cell–cell interactions are governed by membrane surface assemblies. There is no question that membrane function plays a key role in all aspects of cell function.

Biomimetic studies of membranes and membrane transport seek to illustrate some of the fundamental principles governing the complex behaviors of natural systems. Typically this involves greatly simplified systems derived from natural sources [2, 3] or completely artificial systems of even greater simplicity. Indeed, the relationship of the mimic to the natural system is often tenuous. By its very nature a mimic "is applied primarily to artistic or playful imitation, usually suggesting that the copy is ludicrously diminutive or insignificant as compared with the reality imitated" [4]. The biomimetic studies reviewed here are very much in the spirit of this definition.

At the same time there are technological implications of membrane transport. Although pale imitators of natural systems, artificial membrane systems have played an important role in the development of new types of metal ion separations, and new ways to manipulate ion concentrations [5–11]. The systems discussed in this review set out to imitate or illustrate some facet of natural transport systems. In contrast, studies which set out to develop separations or which focus on the properties of specific synthetic carriers will be mentioned only briefly, if at all.

The behavior of naturally occurring low molecular weight transporters in artificial membranes [12–14] will be examined as an aid to developing mimics, but ironically the very object of the mimicry will not be directly examined. Productive generalizations concerning transport proteins can be drawn from recent monographs and reviews [1, 16–17]. These will suffice to develop "ludicrously diminutive" artificial systems which exhibit biomimetic ion transport.

There are two aspects of natural transporter systems which could be imitated: their structure, or their function. Mimics of structure face the daunting problem of the scale of transport proteins, and the relatively low level of structural detail which is available from biochemical studies [16–19]. However, structural mimics of low molecular weight transporters such as the antibiotics gramacidin, valinomycin, nigericin or amphotercin [2, 12, 13] can, and have been studied (Sects. 2 and 4). In contrast, the functional aspects of natural transporters are relatively well known: a specific transport protein mediates a given set of coupled ion fluxes driven by a given set of driving forces. In an artificial mimic of function, the membrane is regarded as a black box. To the extent that the system can reproduce the output (ion flux) from input stimuli (driving forces), the system is apparently biomimetic. As subsequent sections will show, there are numerous ways to produce a desired input/output relationship. The value of biomimetic studies of transport function is to clarify in an artificial context, the constraints placed on natural transport systems. The structural details of these models might be, and usually are, remote from the natural systems they mimic. Ultimately, as the sophistication of the mimics increase, structural and functional aspects will merge.

1.1 Membranes and Transporters

The basic structure of biological membranes is a bilayer of lipid molecules with polar heads exposed to the aqueous phases and the hydrocarbon tails providing a fluid core for the membrane. Transport proteins, and other membrane bound structures are imbedded in the "fluid mosaic" which makes up the functioning membrane [1]. Natural bilayer membranes are highly impermeable to small ions such as K^+ or Cl^- due to the energetic demands for transfer of the ion from water to the hydrocarbon core of the membrane [13]. Dipolar and neutral molecules with favourable water–lipid partition coefficients can traverse biological membranes hence the membrane provides an inherently selective barrier.

In a physico-chemical context, a membrane is "a phase, finite in space, which separates two other phases and exhibits individual resistances to the permeation of different species" [20]. Among other examples, a liquid layer between two gases would constitute a membrane, provided there was some inherent selectivity to the transfer of different gases between the two compartments. Membrane mimetic chemistry, the study of mimics of biological membranes, recognizes numerous types of systems which do not provide a phase barrier [21]. In the context of transport studies however, the phase barrier is essential. Several experimental membranes are in current use.

Bulk liquid membranes consist of an immiscible organic liquid interposed between two aqueous phases held by means of the differences in liquid densities. The simplest form is a U-tube with a dense organic liquid (e.g. chloroform) in the base of the U separating two aqueous phases in the arms [22, 23]. All three phases can be stirred to increase the rate of transport which is controlled by diffusion. Alternative geometries involve cylinders with plane barriers (Schulmann bridges) [24] concentric cylinders [25], density layering in a single test tube [26], and H-tubes (a light organic phase floating on aqueous phases in the base of each vertical arm) [27]. Transport is usually followed by analysis of aliquots withdrawn over time. These systems are relatively unstable to pressure differences and stirring, but are extremely simple to use.

Supported liquid membranes consist of an immiscible organic liquid imbibed into the pores of a porous support, such as filter paper or microporous Teflon, interposed between two aqueous phases [28]. These membranes can be mounted as flat sheets or created in hollow fibre modules [29] to give high surface area/volume ratios. Such membranes are fairly robust and are primarily of industrial interest. *Emulsion liquid membranes* [30, 31] created as a water-in-oil-in-water ternary emulsion are geometrically similar to large vesicles or cells and have some potential interest as biomimetic membranes. This potential has not yet developed and these membranes have been exclusively considered for technological purposes.

Polymer and polymer composite membranes have been widely investigated for applications ranging from ultrafiltration and reverse osmosis to ion selective sensor materials. *Solvent/polymer membranes*, consisting of highly plasticized support polymers are effectively supported liquid membranes with a high viscosity solvent imbibed in the matrix. Their primary application is in ion selective electrodes [20]. *Liquid crystal/polymer composite membranes* are similar in that the plasticizer used is itself a liquid crystal [32]. Membrane permeability is thus subject to thermal control [32, 33]. *Bilayer/polymer composite membranes* are closely similar. The liquid crystalline behaviour is provided by natural lipids or artificial bilayer forming compounds [34] held in a porous support material or as the counter ions to a polyion. Section 3 explores some aspects of biomimetic polymer membranes in more detail.

The most closely biomimetic membranes are formed by suspension of lipids in water. *Vesicles or liposomes*, consisting of a lipid bilayer surrounding an enclosed volume of water offer the bare essentials of a cellular membrane stripped of protein and carbohydrate [21]. Alternatively planar bilayer membranes can be created across a small hole in an inert barrier material such as Teflon. These *black lipid membranes* are particularly useful for the study of the electrical properties of the membrane, and the influence of potential gradients on membrane transport [2, 3, 13, 21]. Section 4 examines bilayer membranes in detail.

The role of transporters in any of these membranes is to catalyse the transport of a substrate across the membrane. If the substrate is ionic and the inherent permeability of the membrane is low, the transporter acts as the sole means for translocation of the substrate. This is defined as facilitated transport. Dipolar or neutral substrates or "leaky" membranes provide a means for the translocation of substrate which is independent of the transporter (non-facilitated transport).

In general, transporters can act by three main strategic mechanisms (Fig. 1). The transporter could form an aqueous pore or channel which spans the membrane. An ionic substrate could then traverse the membrane by diffusion within the channel. Natural transporters utilize this strategy exclusively with numerous variants involving gating, conformational change and activation by small molecules [1]. Alternatively, the transporter could function as a carrier. Transport would occur via a complex of the substrate by the carrier which

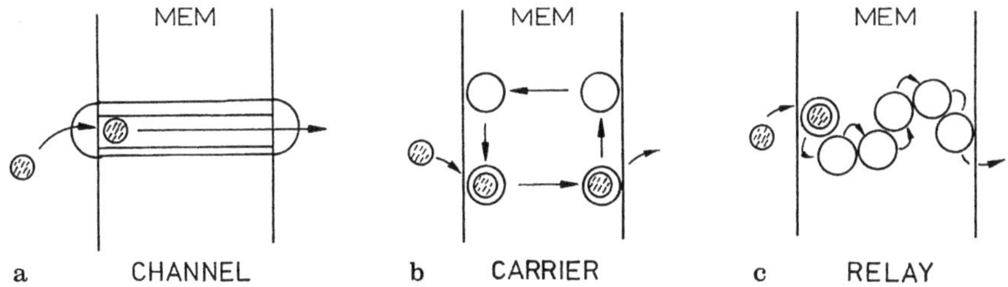

Fig. 1a–c. Strategic mechanisms for transport: channel (**a**), mobile carrier (**b**), relay (**c**)

undergoes free diffusion within the membrane. Some ionophore antibiotics apparently function by this mechanism. The third option is that of a relay. Transport would occur in a series of hops between closely spaced sites within the membrane. Each site stabilizes the ionic substrate relative to the free ion within the membrane. A trivial analogy for the three choices is suggested by a river: a channel is equivalent to a tunnel or bridge, a carrier is equivalent to a boat and a relay is equivalent to a series of stepping stones. Swimming is simply non-facilitated transport across the river.

1.2 Driving Forces and Transport Cycles

Transport of ionic solutes across membranes occurs as a result of some driving force applied to the solute. In general, there are four main driving forces: concentration gradients (more generally gradients in chemical potential), electrical potential gradients, gradients of temperature and gradients of pressure. These forces can produce four main types of flows or fluxes: flow of solute, flow of solvent, flow of charge and flow of heat. Considering the forces and their resultant fluxes pair wise, results in descriptions of the full range of membrane phenomena: osmosis, electrosmosis, Donnan potential, solute transport etc. [20, 35]. The biomimetic emphasis centres on solute transport driven by gradients of chemical potential, although gradients of electrical potential are important especially in bilayer membranes. The flux of some species A might depend on a range of forces and their interactions. A might respond to its own chemical potential gradient $\Delta\mu_A$ or to a gradient in electrical potential $\Delta\phi$. In addition, A might respond to fluxes of other solutes (J_B) or some reaction flux (J_R). In general, the flux of $A(J_A)$ might be given by an equation of the form:

$$J_A = L_{AA}\Delta\mu_A + L_{AE}\Delta\phi + \sum_{B \neq A} L_{AB}J_B + L_{AR}J_R \qquad (1)$$

where the L terms are coefficients which describe how A responds to the driving forces. Equation 1 is completely general but quite uninformative as to the nature of the coefficients or the molecular mechanisms of the transport [20]. Some coefficients have simple forms; for example L_{AA} could be related to diffusion coefficients by Fick's Law. Derivations of the other L coefficients, particularly the couplings to flows of other solutes or reactions, require a more specific understanding of the molecular mechanisms. Several approaches can be taken and in general these all result in closely similar functional forms. Selection of an approach depends solely on the eventual application desired: engineering [36], ion selective electrode behaviour [20], membrane potential [37] or transporter energetics [38, 39]. For the purposes of examining biomimetic ion transport and artificial systems, the approach of Goddard [40] provides a fundamental framework for discussion. Goddard's model is completely general and will simplify to a number of the specific forms derived for the specific applications noted above.

Figures 2–4 illustrate schematic mechanisms for the transport of ionic substrates (A, B) via transporters (L). Although it is simplest to consider L as a mobile carrier, this is not essential: L could equally well be a channel or relay

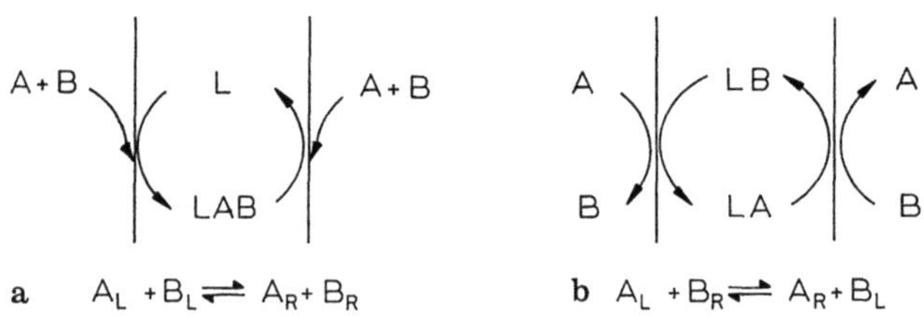

Fig. 2a, b. Schematic mechanism for gradient pumping. Symport (**a**) and antiport (**b**) cycles for the transport of ionic substrates A, B by the duoport transporter L

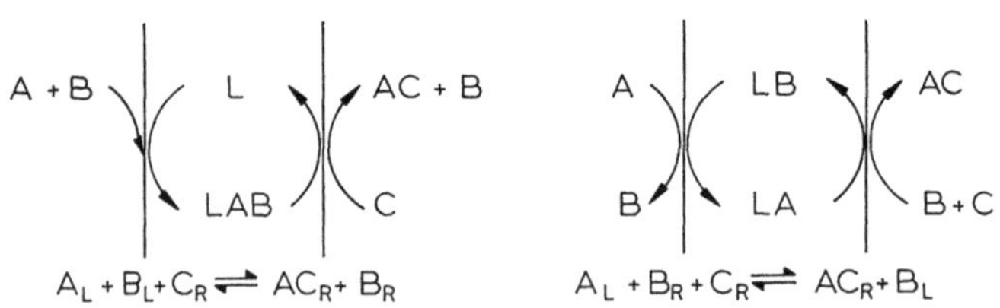

Fig. 3. Schematic mechanism for reaction pumping. Compare to Fig. 2

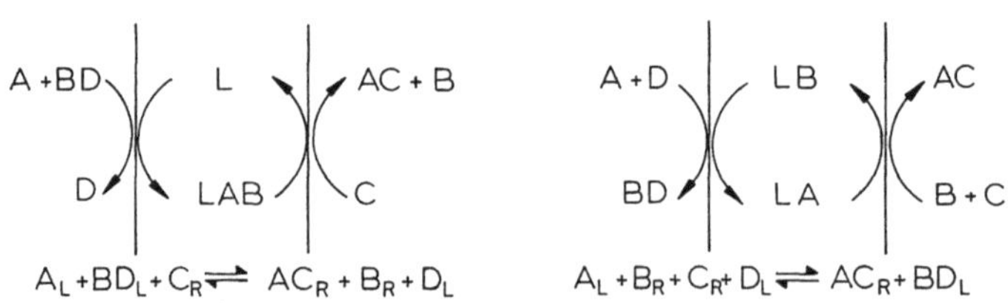

Fig. 4. Schematic mechanism for reaction coupling. Compare to Fig. 3

type transporter. The transporter L is confined to the membrane phase and serves as the sole means of transport of A and B (there is no non-facilitated transport) via the membrane complexes LAB or LA and LB. Since L mediates the transport of both A and B it is a duoport.

Consider the cycle at the left of Fig. 2. At the left aqueous phase interface, A and B are coextracted by the carrier to form LAB. For a mobile carrier, the complex diffuses to the right aqueous interface whereupon the reverse extraction step occurs. Back diffusion of L completes the cycle. Substrates A and B travel in the same direction and their fluxes are coupled. This case is referred to as *symport* or *cotransport*. The equilibrium position of this cycle can be derived from the equilibrium relationship given below the cycle. Since the species A and B in water will be identical in either the left or right aqueous phase, the actual "equilibrium constant" will be unity or:

$$[A_r]/[A_l] = [B_l]/[B_r] \tag{2}$$

If the system at equilibrium is subjected to a change, addition of more A on the right for example, transport will occur to satisfy (2). Clearly this will involve transport of some B from right to left against its concentration gradient. Since the driving force is simply a concentration gradient in A, this is an example of *gradient pumping* or secondary active transport. The cycle illustrated on the right side of Fig. 2 is very similar except that A and B move in opposite directions across the membrane. This is defined as *antiport* or *counter transport*. Again, an imposed concentration gradient in one species can drive the other against its concentration gradient, thus both systems of Fig. 2 can achieve gradient pumping.

Consider now the parallels between Fig. 2 and Figs. 3 and 4. In all cases the cycle on the left results in symport of A and B, while the cycle on the right results in antiport of A and B. The sole differences arise from the additional aqueous phase reactions of A with C (Figs. 3 and 4) and B with D (Fig. 4). Transport in these cycles is coupled to chemical reactions, as well as concentration gradients. Figure 3 represents schematic mechanisms for *primary, active transport* or *reaction pumping* while Fig. 4 illustrates *reaction coupling*. The overall equilibrium positions can be assessed from equilibrium equations derived for each cycle. The actual "equilibrium constant" will not be unity, but will be determined by the free energy of the reaction $A + C \rightarrow AC$ (and $B + D \rightarrow BD$ Fig. 4). The species AC and BD could represent covalently bonded species, or alternatively could be complexes, reduced or oxidized forms (A = electron or hole) or collections of species. The essential point is that a reaction free energy term is available to energize the transport. Reaction pumping systems are developed more fully in Sect. 2.2.

The above discussion of transport cycle energetics is focused on the relationship of the two aqueous phases. Since the transporter is simply a catalyst, its nature plays no role in determining the energetics. The transporter does however, influence the kinetic behaviour of the cycles. Goddard [40] has derived a general relationship between the flux across a membrane and the extraction constants, rates of extraction and rates of translocation of the individual components with the transporter. Provided that the constituent terms can be evaluated for a particular experimental situation, the flux of the cycle can be derived for any system.

The procedure can be illustrated for the symport gradient pumping system of Fig. 5 incorporating crown ether and cryptand carriers for the transport of alkali metal salts across a bulk liquid membrane. This system has been widely investigated by several groups [11, 25, 41–47] and it is apparent that the flux is determined by the diffusion of the carrier and its complexes across the unstirred layers adjacent to the interfaces [41, 42, 48, 49]. The interfacial reactions are fast [48], hence the interfaces are essentially at equilibrium with the bulk phases. The diffusion rates can thus be related to concentration gradients by Fick's Law [40, 41, 48]. The interfacial concentrations can be derived from the extraction equilibrium which applies at either interface.

$$M^+(aq) + X^-(aq) + \bar{L} \rightleftharpoons \overline{LMX} \tag{3}$$

$$K_{ex} = \frac{[\overline{LMX}]}{[\bar{L}][M^+(aq)][X^-(aq)]} \tag{4}$$

The overline indicates the organic phase concentration of the carrier and its complex. The extraction equilibrium constant K_{ex} varies with the carrier–cation association constant, cation–anion ion pairing in the organic phase and the "lipophilicity" of the extracted anion[43, 47, 50].

Simplifying from Goddard's general analysis [40] or proceeding directly from a Fick's Law analysis [48] yields the result that the flux (J) of the cycle of Fig. 5 is given by:

$$J = \frac{DL_o K_{ex}}{2l} \frac{([M^+]_l[X^-]_l - [M^+]_r[X^-]_r)}{(1 + K_{ex}[M^+]_l[X^-]_l)(1 + K_{ex}[M^+]_r[X^-]_r)} \tag{5}$$

D is a diffusion coefficient of L and LMX in the membrane, l is the thickness of the unstirred layer and L_o is the total carrier concentration. The term $DL_o/2l$ is the maximum flux the system can exhibit (J_{max}) [47, 48] and gives an alternate form of (5) in terms of a normalized flux (J/J_{max}):

$$\frac{J}{J\,max} = K_{ex} \frac{([M^+]_l[X^-]_l - [M^+]_r[X^-]_r)}{(1 + K_{ex}[M^+]_l[X^-]_l)(1 + K_{ex}[M^+]_r[X^-]_r)} \tag{6}$$

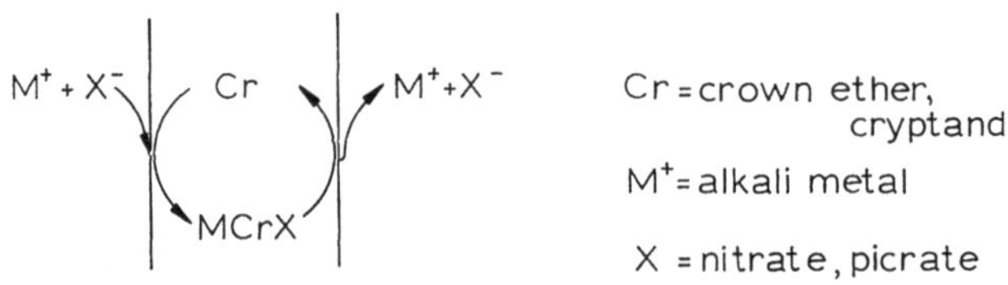

Fig. 5. Symport gradient pumping by mobile carriers

For small values of K_{ex}, the denominator terms of Eq. 6 are negligible and the normalized flux will increase as K_{ex} to a maximum at some optimal value of the extraction constant (K_{ex}(opt)). This increase behaves as a saturable process and rearranged versions of equation 6 have the form of a Michealis-Menton expression [40, 48, 50].

At the K_{ex}(opt) the carrier is half saturated, that is $\overline{[L]} = \overline{[LMX]}$ in the membrane. A further increase in K_{ex} results in a decrease in flux as the amount of free L available for the return cycle diminishes. As illustrated in Fig. 6, the normalized flux is thus a bell shaped function of log K_{ex}. The theoretical curve can be computed from Eq. 6 for any set of initial conditions [48, 50] and compared with experiment. Figure 6 shows two examples for two different sets of experimental conditions [44, 47]. The required extraction constants were measured directly in one case [47], but estimated from the cation–carrier association constants in the other [50]. The agreement of theory and experiment is adequate: the scatter is primarily due to the loose control of the experimental variables. The key point is that the maximum of the two curves (log K_{ex}(opt)) is different in the two cases as is generally expected [48]. The difference in breadth of the two curves is also expected: higher source phase concentrations lead to lower values of K_{ex}(opt) and broader peak maxima in symport systems [48].

It is important to emphasize that the apparent behaviour of this system, and all other transport systems, is strongly dependent upon the experimental conditions. A carrier/salt combination with a log $K_{ex} = 3$ would appear to be mediocre under one set of conditions of Fig. 6, while it would be close to optimal for the other set of conditions. The behavior of antiport systems is similar except that K_{ex}(opt) is always unity for symmetrical cases [48]. An antiport system is most efficient when the transporter has equal affinity for the two substrates. Peak breadth, however, increases with the source phase concentration [48].

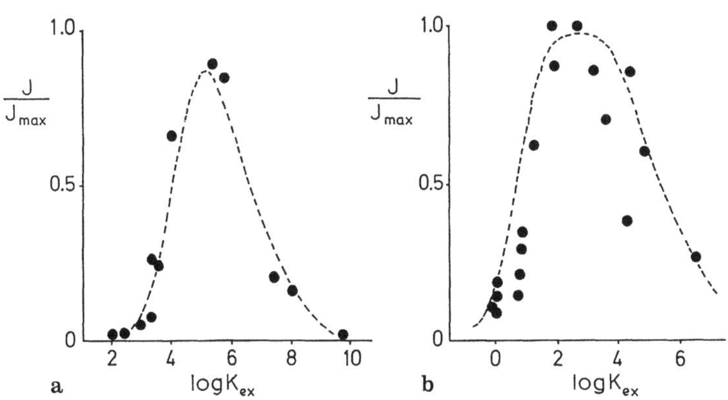

Fig. 6a, b. Normalized flux (J/J_{max}) as a function of log K_{ex} for two systems related to Fig. 5. Alkali metal picrate transport by cryptand carriers (**a**) derived from [47]. Alkali metal salt transport by crown ether carriers (**b**) ([50] using data of [43, 44])

Transporter selectivity is also subject to control by experimental conditions. The selectivity of transport is defined as the ratio of fluxes for substrates A and B in competition (J_A/J_B). There will be two extraction constants $K_{ex}(A)$ and $K_{ex}(B)$ and three sets of relationships with $K_{ex}(opt)$. Assume that A is selectively extracted relative to B, i.e. that $K_{ex}(A) > K_{ex}(B)$. If both $K_{ex}(A)$ and $K_{ex}(B)$ lie below $K_{ex}(opt)$ then A will be extracted and transported more rapidly than B and the membrane selectivity will reflect the inherent extraction selectivity. If both $K_{ex}(A)$ and $K_{ex}(B)$ lie above $K_{ex}(opt)$, B will be preferentially released and the reverse selectivity will be observed. Finally if $K_{ex}(B)$ lies below and $K_{ex}(A)$ lies above $K_{ex}(opt)$, there might well be a completely unselective behavior of the system. The flux, hence the selectivity will depend on extent of transport, as well as on the initial conditions which define the J/J max vs log K_{ex} bell curve [48]. In no case will there be a simple relationship between the fluxes observed for the substrates in competition when compared with fluxes obtained for each substrate singly.

Head-to-head comparisons between synthetic and natural transporters are thus fraught with difficulties. Even comparisons between synthetic transporters – their ability to achieve transport, their "inherent" selectivities – will be difficult. This is not simply an artifact of the particular derivation above for bulk liquid membranes, but is a fundamental constraint on transporters in general. The phrase "transporter X is better than transporter Y" is virtually without meaning unless extensively qualified.

2 Mobile Carriers

Small molecules are easier to make than large molecules and offer greater control over their structures. As a consequence, mobile carriers have been widely investigated for biomimetic ion transport even though naturally occurring mobile carriers play a relatively minor role in the general scheme of biological ion transport. Ionophore antibiotics, derived from natural sources [51–53], appear to be secondary metabolites and play no direct ion transport role in the organisms which produce them. As illustrated in Fig. 7, a number of different structural types can be identified.

The structures on the top line of Fig. 7 are neutral mobile carriers and in general produce symport of cations and anions in biological and artificial membranes [51]. In contrast, the remaining structures on Fig. 7 are carboxylic acid ionophores capable of antiport of cations and protons [54]. The critical feature of all ionophore antibiotics, is their ability to form inclusion complexes of cations. Examination of solid state structures [53] shows ionophore – cation complexes with the cation lying near the centre of the ligand, bound by 6–8 donor oxygen atoms. The ligand is wrapped around the cation and presents an exterior face covered in alkyl and aryl groups. As a result, the complexes of the ionophores of Fig. 7 are all highly soluble in organic solvents, but essentially

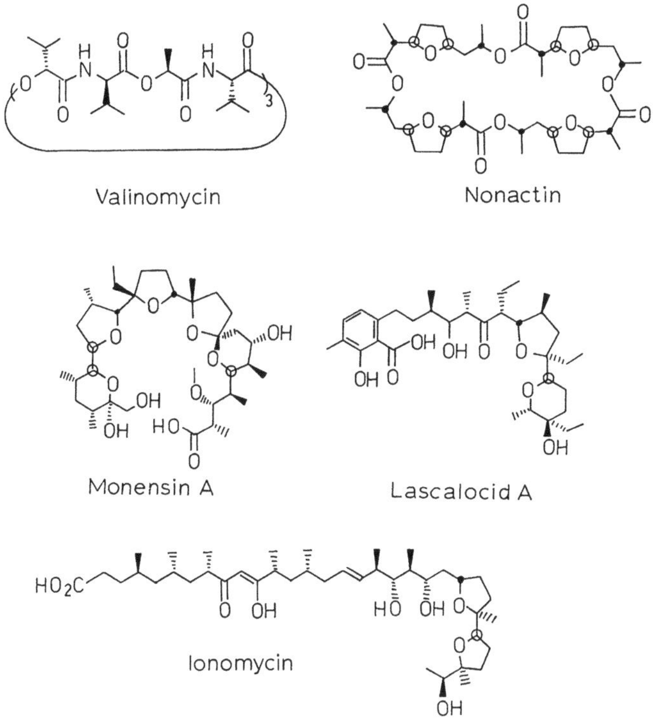

Fig. 7. Naturally occurring mobile carrier ionophores

insoluble in aqueous solution. Partitioning of the cation from an aqueous to an organic phase is thus highly favourable.

The transport properties of these ionophores have been widely studied in artificial membrane systems [12–14, 51, 54–58]: bulk liquid membranes, supported liquid membranes, black lipid membranes and vesicles. The transport mechanisms in liquid membranes are similar to those outlined in Sect. 1.3. In vesicle systems, additional associations with lipid components and specific interfacial effects introduce additional complexity [54]. NMR provides a powerful means of monitoring transport in vesicle systems, either by monitoring the effects of transport of paramagnetic ions on the lipid head group resonances [59], or by direct dynamic NMR studies of the transported ions [60, 61]. To date, these types of studies have not generally extended to synthetic carriers (but see ref. [62]). The naturally occurring ionophores can be induced to transport "unusual" substrates (transition metal ions and ammonium salts) [63] and modifications to their structures provide new types of semi-synthetic ionophores with altered properties [64].

The extent of direct *structural* imitation of naturally occurring ionophore antibiotics depends on the readers tolerance and ability to "suspend disbelief".

Functional mimics clearly should possess an ion binding region, suitable lipophilic properties to achieve partitioning of the complexes to organic phases, and for carboxylate ionophores, a site for ionization. The extent to which these general functions are satisfied by specific structural features acting individually is not clear from the ionophores themselves, hence different workers view the problem of mimic design differently. Some examples of "structural" mimics are illustrated in Fig. 8, (neutral ionophores) and Fig. 9 (carboxylate ionophores).

Nactin mimics 8A/8B and related structures [65] possess a 32 membered ring containing 4 ester units, as does the natural product. For ease of synthesis, however, the model has a mirror plane of symmetry which is absent in the parent. The mimics can extract alkali metal picrates into chloroform with relatively low extraction constant and transport metal picrates slowly through a bulk liquid membrane. Clearly the ion binding site is not optimal. Although not explicitly designed as valinomycin mimics, macrocyclic lactams such as 8E have been studied extensively by Shanzer [68]. The overall symmetry of the macro-

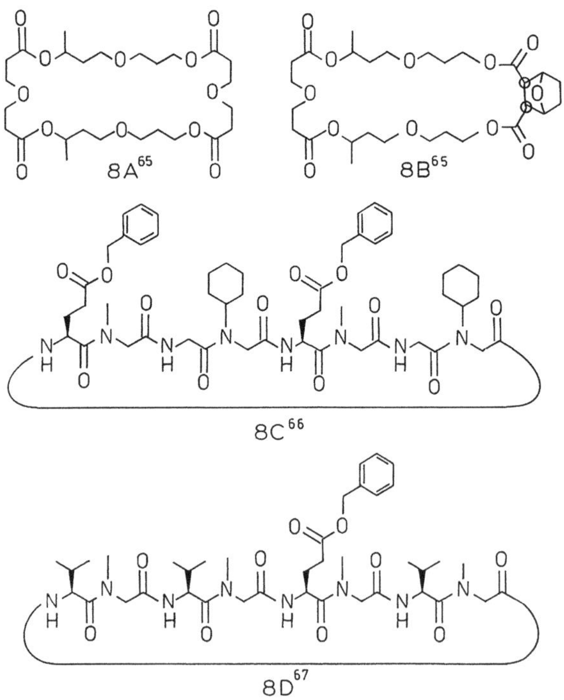

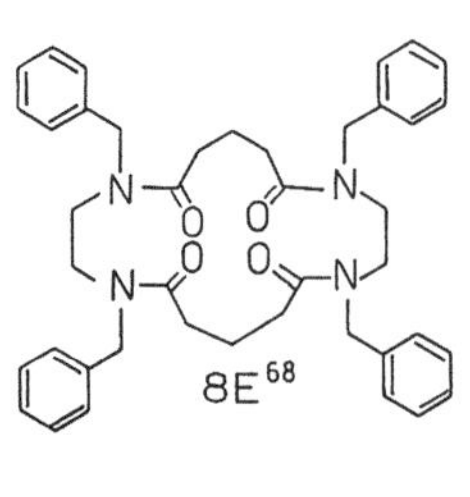

Fig. 8. "Structural" mimics of neutral cyclic ionophore antibiotics [65–68]

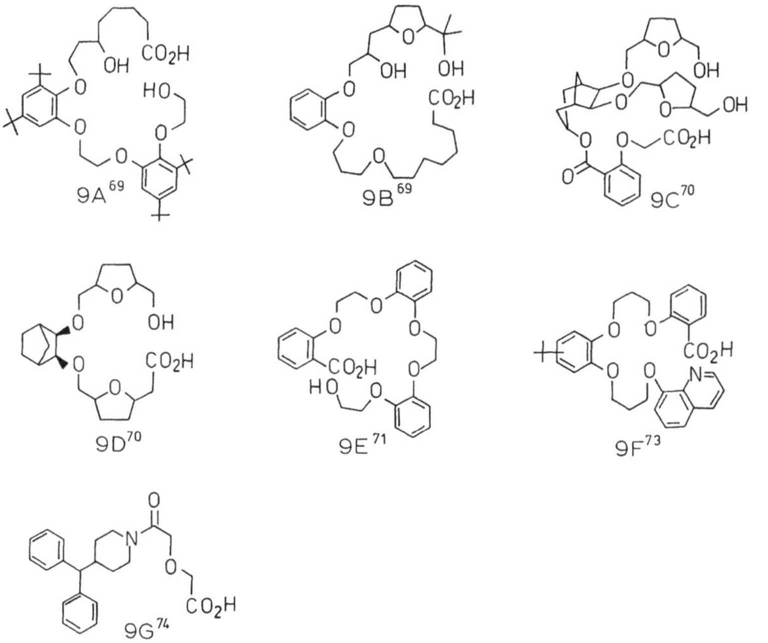

Fig. 9. "Structural" mimics of acyclic carboxylate ionophore antibiotics [69–74]

cycle and the conformations of the free ligand can be controlled to create efficient Li^+ ionophores. A similar approach may be applicable to the larger ring materials as well.

Cyclic octapeptides such as 8C [66] and 8D [67], are loose mimics of depsipeptide carriers such as valinomycin. These materials were designed as Ca^{2+} ionophores and in that sense are functional mimics of the antibiotic beauvericin [75]. The cyclic octapeptide 8C transports Ca^{2+} across bulk liquid membanes with high selectivity for Ca^{2+} in competition with alkali metal and alkaline earth cations [66]. It also transports Ca^{2+} across bilayer membranes in vesicles. In both systems transport is anion dependent: the anion is carried as a symport species in bulk liquid membrane or as both symport and antiport in vesicles [66]. The last result implies transport driven by transmembrane potential can occur. The cyclic octapeptide 8D transports Ca^{2+} across vesicles, but the anion dependence has not been reported [67].

"Structural" mimics of the carboxylate antibiotics shown in Fig. 9 bear much less direct relationships to their parents than the mimics of neutral ionophores of Fig. 8. Compounds such as 9A and 9B contain a single carboxylic acid, five or six ether and hydroxyl groups for cation binding and similar overall size and lipophilicity when compared to monensin [69]. This was insufficient to produce biologically active materials but the actual transport abilities of these materials were not reported [69]. Tetrahydrofuran containing mimics 9C and 9D [70] were capable of transporting Ca^{2+} across bulk liquid membranes at rates comparable to antibiotic A23187. The stoichiometry of the cycle was not reported but the direction of transport was consistent with Ca^{2+}-proton antiport [69]. This would require a stoichiometry of two carriers per metal ion. The acyclic polyether acids 9E [71, 72] and 9F [73] were not designed as structural mimics, but exhibit encapsulation of the bound cation by a coiled or pseudocyclic array. Carrier 9E selectively transports K^+ via an antiport with protons, from a mixture of Na^+ and K^+ [72]; 9F does the same for Li^+ from a mixture of alkali metal cations [73]. "Structural" mimics such as 9G [74] might better be regarded as functional mimics; 9G transports Ca^{2+} at rates comparable to A23187 via a cation–proton antiport of unreported stoichiometry.

2.1 Functional Mimics–Gradient Pumping

2.1.1 Cation–Proton Antiport

Schematic mechanisms for cation–proton antiport of mono and divalent ions are given in Fig. 10. The left aqueous phase is typically basic with respect to the right aqueous phase thus the cycle 'turns" counter clockwise to result in the transfer of metal ions from left to right. Cycle A results in transfer of monovalent ions, as is commonly observed for carboxylate ionophore antibiotics such as monensin [12, 54, 57]. Two-to-one complexes of ionophore-divalent cation occur in transport cycles involving ionophores such as lascalocid and A23187 (calcimycin) [54, 57]. Cycle C involves a diacid carrier to achieve divalent ion transport via a 1:1 complex. A cycle of this stoichiometry applies to ionomycin [76], a carboxylate antibiotic with a second ionizable group (β-diketone). Cycles involving symport of divalent ions with monovalent anions, such as D, are also possible [54].

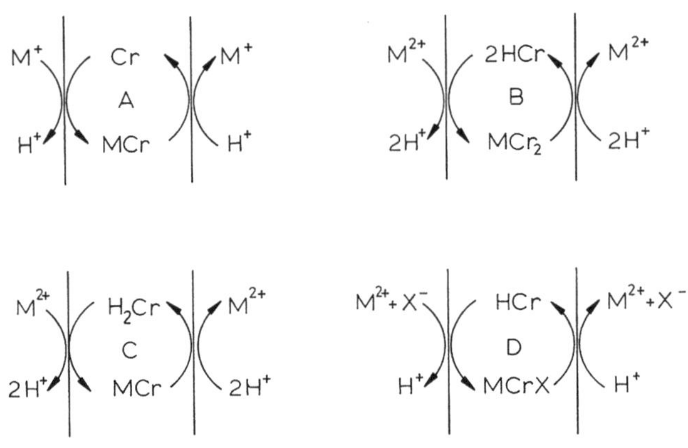

Fig. 10. Schematic mechanisms for cation–proton antiport of alkali metal and alkaline earth cations

All the cycles of Fig. 10 have been examined in liquid membranes using the crown ether carriers illustrated in Fig. 11 [23, 77–82]. Not surprisingly, the cycle stoichiometries expected were confirmed in all cases [23, 78]. Transport was highly efficient (one proton drives one alkali metal cation) but "leakage" could be induced by addition of lipophilic anions or lipophilic cations into the system [23]. The balance between cycles B and D is controlled by the nature of the anions in the left aqueous (basic phase). In the absence of easily extracted anions, cycle B is observed. The 2:1 complex in the organic phase involves the metal ion bound in one carrier crown ether with the second carboxylate crown ether acting only as a counterion [79]. The presence of very liposoluble anions such as picrate shifts the cycle to D; the "empty" crown ether has simply been replaced by another anion [78, 79]; less easily extracted anions are unable to provoke a shift from cycle B to D.

The selectivity of these carriers in membrane transport systems is rather modest although monovalent/divalent selectivity is high. For dicarboxylate carriers, this result is not unexpected since only a single cation binding site is

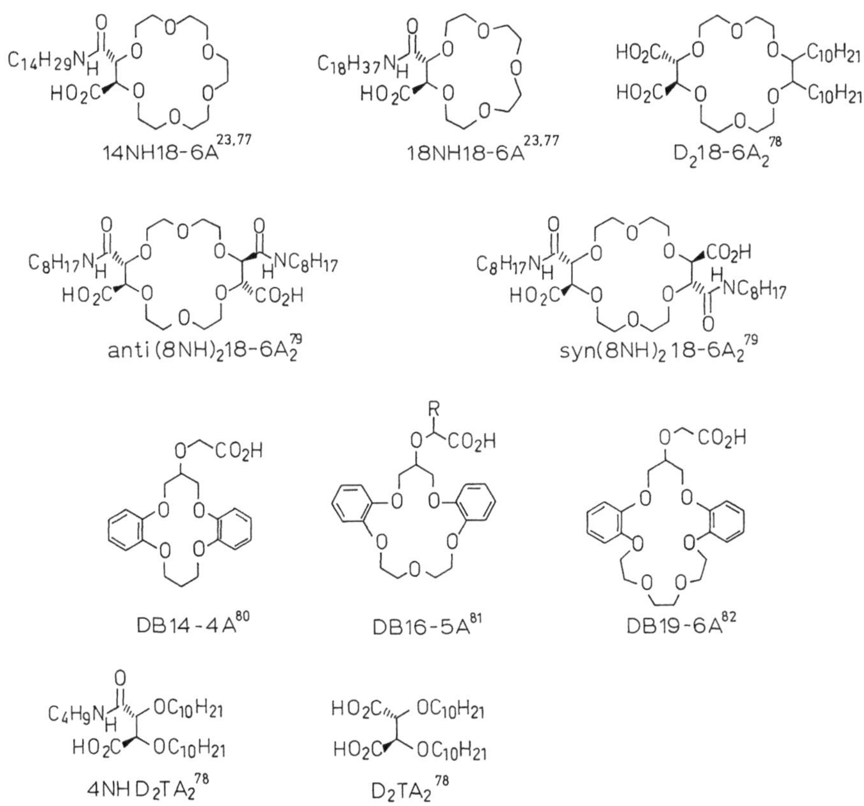

Fig. 11. Synthetic carboxylate ionophores for antiport cation–proton gradient pumping [23, 77–82]

provided. The high divalent selectivity of monocarboxylate carriers is more difficult to understand as there appears to be no reason why cycle B should be greatly favoured over cycle C. Figure 12 illustrates the schematic transport cycle for competitive transport of K^+ and Sr^{2+} and the experimental concentration–time curves for the system. It is clear that K^+ transport is inhibited by the transport of Sr^{2+} until the Sr^{2+} concentration falls to some low level [83].

The system has been examined in considerable detail and exhibits the following characteristics: i) The kinetic order of the transport reaction is a function of the initial conditions. At high substrate/carrier ratios it is zero order in carrier, becoming first order in a carrier–cation complex at lower ratios [23, 77]. The system of Fig. 12 is in the first order kinetics regime for both cations. ii) The transport in both kinetic regimes is diffusion limited at low stirring rates, but becomes independent of stirring rate above 200 rpm (as in Fig. 12). This indicates that interfacial processes control the transport under the conditions of Fig. 12 [22, 83]. From the kinetic orders [23, 77, 78], interfacial surface activity of the carriers [77, 78] and the stirring results [23], the transport is apparently limited by adsorption–desorption processes at the aqueous–organic interface. In the proposed mechanism, the carriers form an interfacial layer which is in equilibrium with the aqueous phase. On the basic side, the interfacial region establishes a negative surface potential by ionization of the carboxylates which implies an equilibrium double layer is also established. Cation binding into a Stern layer of crown ethers completes the interfacial region [83]. Applying this kinetic/equilibrium model to the case of Fig. 12 reveals that the high Sr^{2+} selectivity is composed of three contributions: an inherent cation binding selectivity favouring K^+ by a factor of 4, a kinetic factor favouring desorption of Sr^{2+} complexes by a factor of 13, and the double-layer electrostatic "selectivity" favouring Sr^{2+} by a factor of 80 [83]. Thus one of the

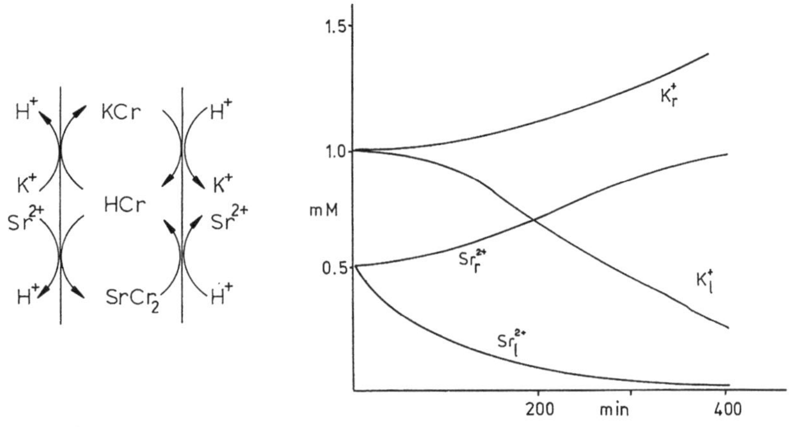

Fig. 12. Competitive K^+/Sr^{2+} transport by a monocarboxylate 18-crown-6 carrier [83]

reasons why the system is divalent selective is simply because divalent ions are enriched at the interface where extraction occurs. Similar effects must occur at charged biological membrane surfaces.

The structures included in Fig. 11 also include two simple tartaric acid diethers. To the chagrin of synthetic crown ether chemists, these simple acids act as relatively efficient transporters of divalent ions [78]. Apparently all that is required for a functional mimic is a suitable balance of lipophilic groups, a carboxylic acid and a few well placed oxygens. Stearic acid is inactive as a carrier as is a mixture of dicyclohexyl 18-crown-6 and stearic acid at a 1 mM concentration [23]. Higher concentrations of this carrier mixture and higher source phase concentrations of cations act together to produce a functional transport system [84] presumably by increasing the breadth of the J/J_{max} vs log K_{ex} bell curve to include the K_{ex} value for this system.

The common configuration for cation–proton antiport systems is with a large imposed proton gradient between phases of equal metal ion concentration. Collapse of the proton gradient drives the metal ion to create a concentration gradient in the cation. The reverse can also occur; an imposed cation gradient will create a pH gradient using monensin as a carrier [56]. The system thus behaves as implied in the reversible cycles of Fig. 10. A modification of the basic cycle in which enforced release of metal ion occurs is given in Fig. 13 [85, 86]. The carrier exhibits pK_a's at 6.99 (NH^+) and 3.25 (CO_2H). The zwitterionic form is maximal at pH = 5.2. Transport is most efficient when the left phase is strongly basic, to deprotonate the carrier to its 1^- form, and when the right phase is close to pH 5.2. More strongly acidic solutions protonate on both carboxylate and nitrogen and the resulting cationic carrier cannot fulfil the return cycle. At the same time, N-protonation is required to assist in loss of metal to the right hand aqueous phase [85] and transport will occur only slowly to a pH 7 receiving a solution.

2.1.2 Cation–Anion Symport

Cation–anion symport driven by concentration gradients alone is relatively rare in a biochemical context. Valinomycin and the other neutral ionophores of Fig. 7 will transport metal salts across bulk liquid membranes by this process [12] but in bilayer membranes, transport is more usually achieved by charged complexes driven by potential gradients [13, 14, 37]. There have been relatively

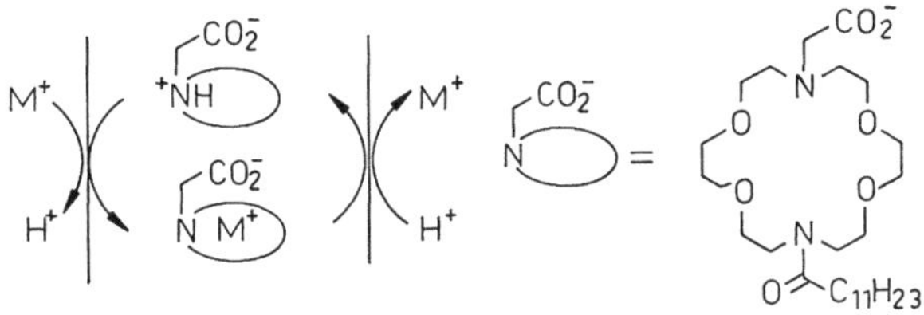

Fig. 13. "Enforced" ion release by a zwitterionic crown ether carrier [85, 86]

few studies of synthetic carrier ionophores in vesicle membranes [62, 66, 87, 88] where contributions by potential gradients might be observed. These cases examined required the presence of an "uncoupler", a weak acid, to accelerate proton antiport and thus collapse any potential gradient created [88]. A crown ether carboxylic acid related to the structures of Fig. 11 mediates direct cation–proton antiport without the need for additional uncoupler [87]. Calcium transport by a cyclo-octapeptide [66] was shown to be anion dependent, consistent with at least some contribution from cation–anion symport cycles.

The major interest in cation–anion symport studies in bulk liquid membranes centres on the range of substrates which can be transported, and on control of selectivity through carrier structure. Metal ion transport has been widely investigated [9, 10, 41–48], but is of relatively minor interest in a biomimetic context due to inherent mechanistic differences between bilayers and bulk liquid membranes as noted above. Of greater interest are non metal cations; simple ammonium salts [89], salts of phenylethylamine and its derivatives [90, 91] and salts of amino acid esters [89, 92–97]. Some systems are summarized in Fig. 14.

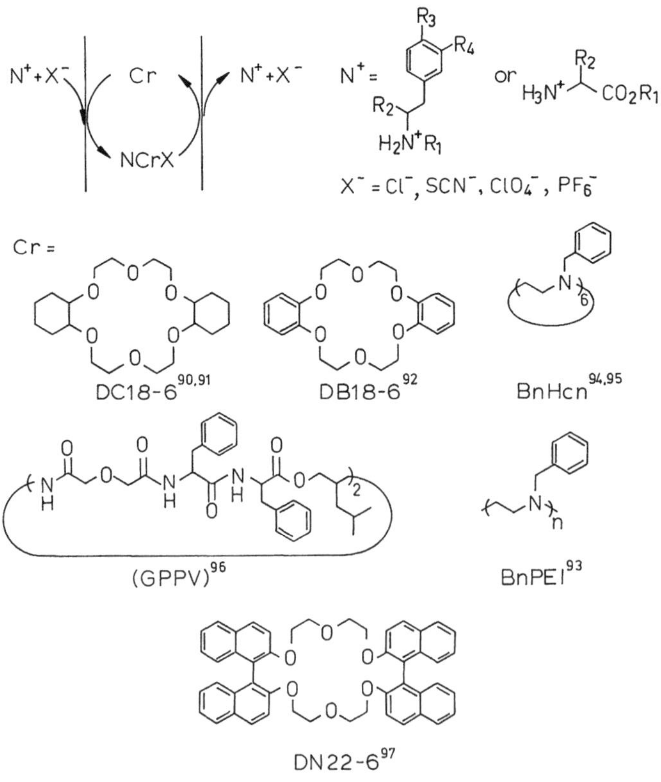

Fig. 14. Ammonium salt transport systems [90–97]

Crown ethers bind primary ammonium ions by acting as hydrogen bond acceptors from the NH_3^+. Thus biogenic phenylethylamine derivatives could be transported by dicyclohexyl 18-crown-6 [90, 91] and amino acid esters by dibenzo 18-crown-6 [92] and other related oxa-crown ethers [95]. Since the transporter in these cases can also carry alkali metal cations, ammonium ion transport is subject to competitive inhibition [90–92] by alkali metal cations. Primary ammonium salts are transported, but secondary ammonium salts (ephedrine) or very hindered primary (phentermine) are not transported [90]. Hydrogen bonding from NH_3^+ to nitrogen is also possible and poly-aza-macrocycles [94, 95] and linear polyamines [93] act as carriers of amino acid ester salts. Competitive inhibition by alkali metals is much less in these cases due to the lower affinity of the carrier for these cations [94]. The lipophilicity of the cotransported anion [93–95] and substrate [89–92] both control the transport rate and selectivity.

The ammonium salts utilized are chiral in many cases as are some of the carriers [96, 97]. The possibility thus exists for a chiral recognition in the extraction step leading to preferential extraction and transport of one enantiomer from a racemic mixture. This potential has been elegantly realized as shown in Fig. 15 [97]; a resolving machine for phenylglycine methyl ester salts. This system employs a W shaped cell containing two separate liquid membranes in the base of each V with a common source phase at the center. The left membrane contains the *SS*-binaphthylcrown carrier (Fig. 14) while the right membrane contains the enantiomeric *RR*-carrier. Interaction of a given carrier with racemic phenylglycine methyl ester (RG^+ and SG^+) results in a pair of diastereomeric complexes. These are of unequal stability with the net result that one enantiomer of the guest is preferentially extracted. From CPK models, it is apparent that the *R*,*R*-carrier will preferentially extract the *R*-guest [97]. In the complete system (Fig. 15) the major cycles operate to move the *S*-guest to the left and the *R*-guest to the right. For small amounts of transport, guests of 90% optical purity were obtained in the outer arms of the cell [97]. Since the system at equilibrium will be independent of the membrane composition, the resolution

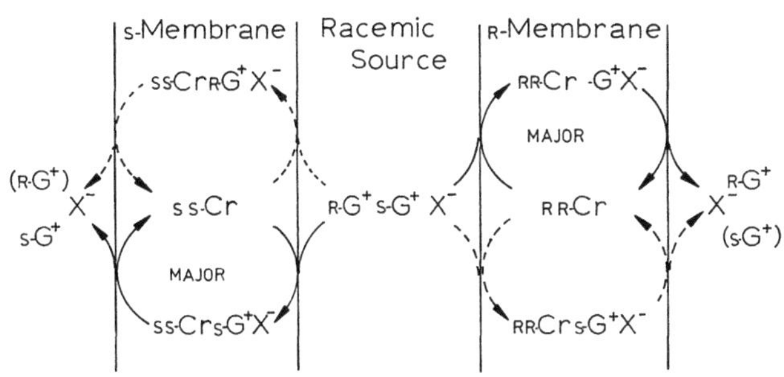

Fig. 15. Schematic mechanism of a "resolving machine" for chiral ammonium salts [97]

must degrade as transport continues. Even so, this system represents a simple model for chiral recognition in biological systems.

2.1.3 Anion Transport Systems

Biomimetic studies of anion transport, using simple anion exchangers as carriers, actually predate the large number of cation transport studies which now dominate the filed [98, 99]. Part of the emphasis on cation transport is due to the ionophore antibiotics: no naturally occurring anion ionophores have been isolated. Anion transport in natural systems is widely evident [1], but is so far restricted to transport proteins as the natural transporters. Another consideration is the greater emphasis traditionally placed on cation coordination when compared with the coordination chemistry of anions. The range of available anion complexing agents is thus smaller, and the total number of examples is vastly less. Even so, anion transport systems have received some attention.

Figure 16 shows some examples of anion–anion antiport and anion–cation symport systems in which the anion is extracted as the counter ion to some lipophilic cationic species. These include lipophilic transition metal complexes [100, 101], where the metal remains fixed in the membrane, as well as ionophore–alkali metal complexes where cation symport results in loss of metal from the membrane phase [102–104]. Careful control of pH permits the system of Fig. 13 to be used as an anion transport system for amino acids [66]; the imposed pH gradient is sufficient to enforce release of the anion, but insufficient to release the Ca^{2+} bound in the crown ether. Bis-quaternary ammonium salts act as carriers of dianionic substrates [105] and can be used as selective extractants of the dianionic forms of ADP and ATP [106]. As well, protonated polyamines can act as carriers with a complex range of cycle stoichiometries [107–109].

In all these cases the anion is held in association with the carrier by ion pairing interactions. The "selectivity" of extraction and transport is predominantly governed by the lipophilicity of the anionic substrate with a strong tendency to favour the more lipophilic anions. The carrier itself exerts only a minor role in discriminating between competing substrates. An alternative approach is sketched in Fig. 17 in which the anion is transferred via inclusion complexes of a macrobicyclic carrier [109]. The selectivity of inclusion complexation is governed by the complementary of shape, size and functionality between the anionic substrate and the ligand [110–112]. However, all the anion inclusion complexes known possess an excess of positive charge. As shown in Fig. 17, addition of a large, liposoluble anion (DNNS) is required to neutralize the excess charge. Since the DNNS is too large to penetrate the carrier, it allows a single substrate anion to be transferred. The systems examined are highly Br^- selective when compared to NO_3^-, an anion of similar lipophilicity [109]. The system is still not ideal as more lipophilic anions such as ClO_4^- still compete even though they are probably not bound within the ligand cavity [109].

Even carriers based on electrostatic interactions can exhibit some subtle effects. Figure 18 illustrates a chiral recognition in the transport of mandalate

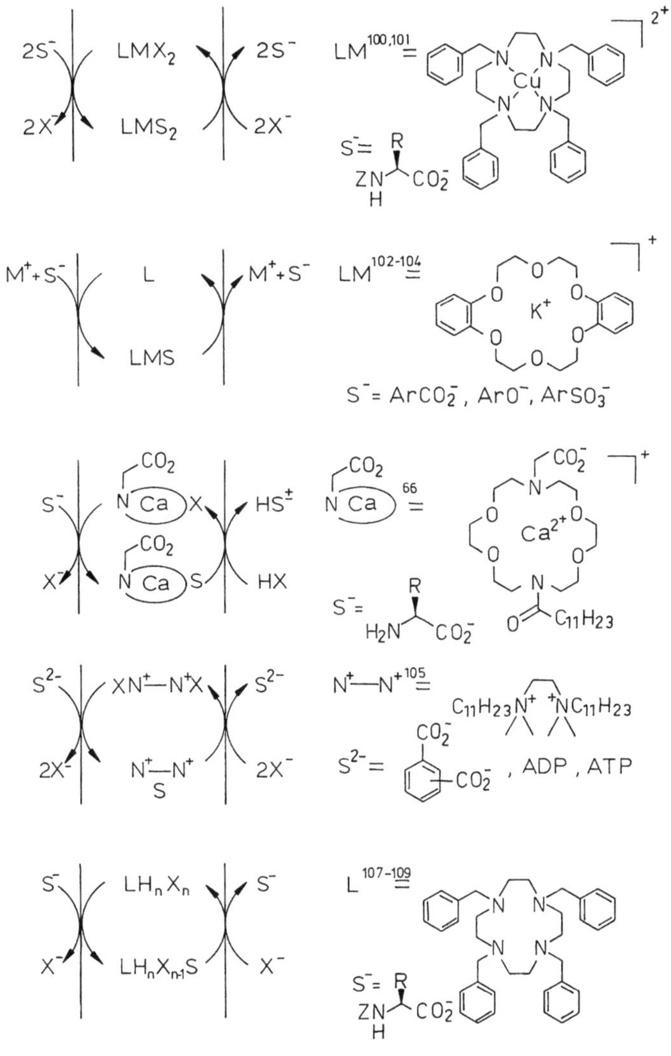

Fig. 16. Ion pairing anion transport systems [100–108]

anions by a simple chiral ammonium salt [113]. The system is an anion–anion antiport and the overall rate is optimal for propionate or butyrate as the counter transported X^-. The salt complexes in the membrane are diastereomeric and should have different extraction constants; in fact the (+)-mandelate isomer is preferentially exchanged for all X^- examined. The extent of this selectivity is a function of X^- (propionate is optimal) as well as a function of ionic strength. Thus a system constructed with a low ionic strength receiving phase initially transports (+)-mandelate selectively, with the extent of chiral resolution de-

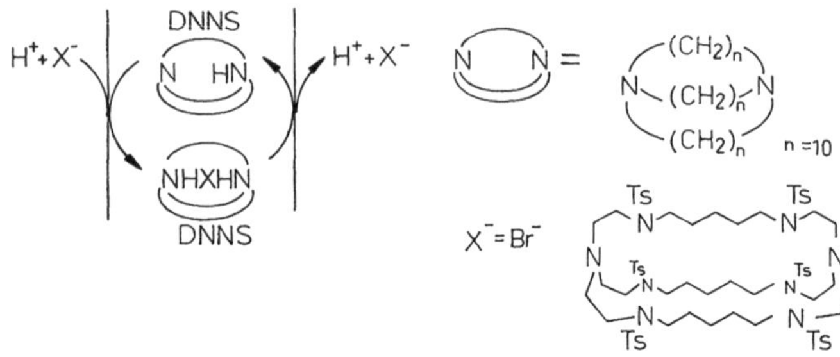

Fig. 17. Anion–proton symport mediated by anion inclusion complexes [109]

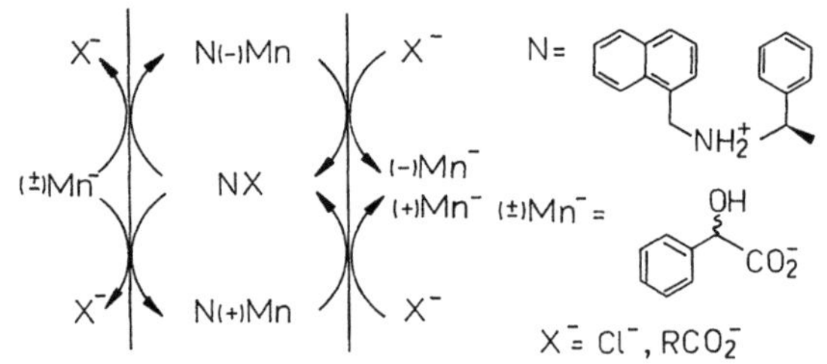

Fig. 18. Chiral selectivity in mandelate ion transport [113]

pendent upon the antiport anion X^- [113]. At a very primitive level, this represents a *regulation* of *chiral resolution* (see Sect. 2.3).

Another intriguing anion transport system is illustrated in Fig. 19 [114]; the interest in this system is not on the anions but on the co-transfer of monosaccharides. At the basic interface, phenyl boronic acid forms an anionic complex with monosaccharides containing *cis*-diol units (fructose, maltose). The counter cation is trioctylmethylammonium, hence the whole complex can be extracted. At the acidic interface, decomposition of the boronate complex releases the monosaccharide to the aqueous phase. The system will pump sugars against a concentration gradient and is somewhat selective for monosaccharides containing *cis*-diol units. Transport systems for hydrophilic neutrals (urea) or zwitterionic species have not yet received much attention. Transport systems for hydrophobic neutral substrates involve "reverse" polarity membranes (an aqueous membrane interposed between two hydrocarbon layers) [115–117]. These are of some technological interest.

2.2. Functional Mimics–Reaction Pumping and Reaction Coupling

The schematic mechanisms for reaction pumping and reaction coupling given in Figs. 3 and 4 provide little insight into the chemical identities of the species. As a consequence, these types of functional mimics have been much less investigated. Figures 20–23 present the unambiguous cases which have appeared to date. Numerous other cases of "uphill" or "active" transport have been reported, but

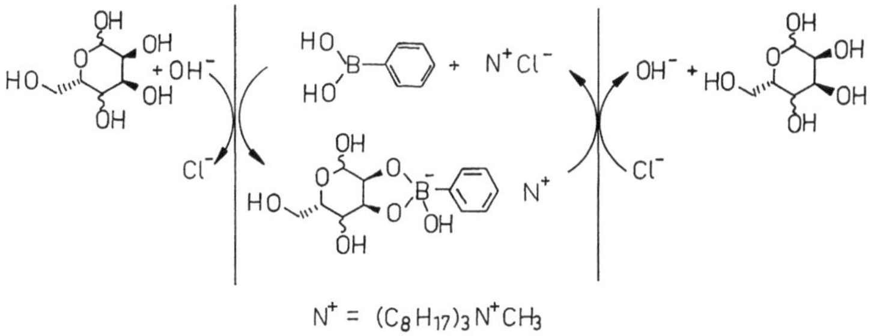

Fig. 19. Monosaccharide transport via boronate complexes [114]

a closer analysis reveals the majority to be examples of gradient pumping [127–129], or more interestingly, examples of photo- or redox-switching (Sect. 2.3) energized by gradient pumping.

The problem of "active" transport is most easily envisaged in terms of redox cycles in which transfer of electrons across a membrane couples two separated half cells (Fig. 20). Direct stoichiometric cycles [118] or a photo-catalytic cycle involving a similar membrane system have been reported [119]. The pH gradient created by these cycles was not quantified, but the overall behaviour observed is consistent with cycles illustrated. Coupling of electron transfer to cation symport (Fig. 21) or anion antiport (Fig. 22) requires additional carriers for the ionic species. One strategy uses co-carriers: a carrier for the cation or anion and a second redox-active carrier for the electron [120, 123]. Altern-

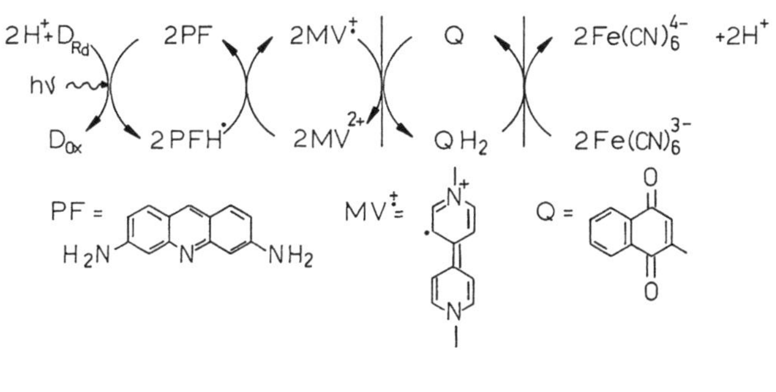

Fig. 20. Electron–cation symport reaction coupling [118, 119]

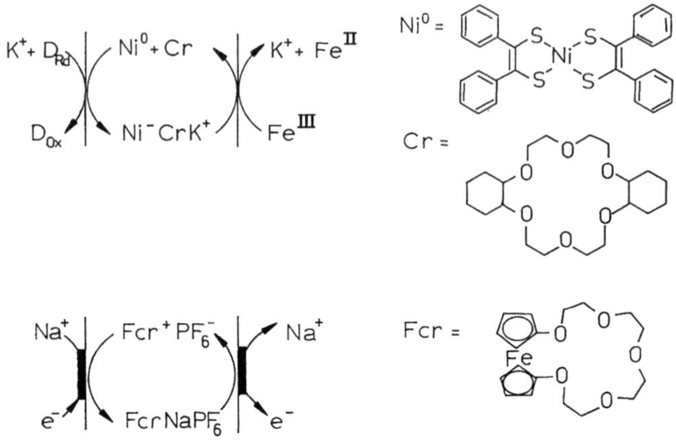

Fig. 21. Electron–cation symport reaction pumping [120, 121]

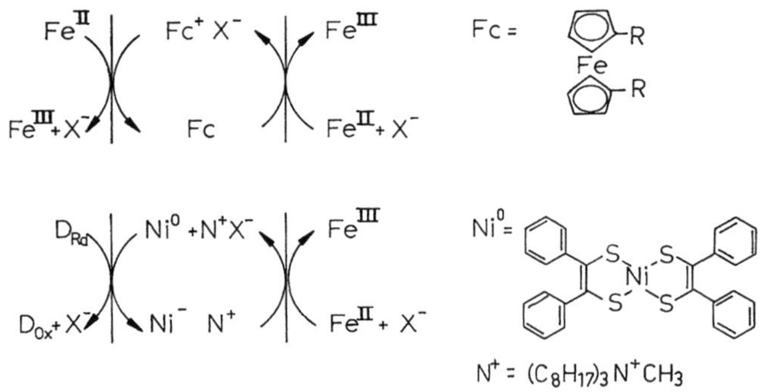

Fig. 22. Electron–anion antiport reaction pumping [122, 123]

atively, the redox-active species can act as the carrier, either as a liposoluble counterion [122] or as a specifically designed ferrocene crown ether [121]. In this latter system (bottom Fig. 21), the electron source and sink were Pt minigrids, rather than the chemical oxidants and reductants used in the other cases.

"Active" transport systems which do not involve electron transfer are illustrated in Fig. 23. These are two very different systems which derive the driving force for the cycle from two different sources. In the upper system, conventional cation–anion symport using cryptand carriers is energized by formation of a complex of the transported cation in the right hand aqueous phase [124]. The kinetic behavior and the equilibrium position of this cycle could be predicted from the Goddard analysis [40, 124]. The 1:1 association constant in the right aqueous phase provides a free energy change of 24 kJ to

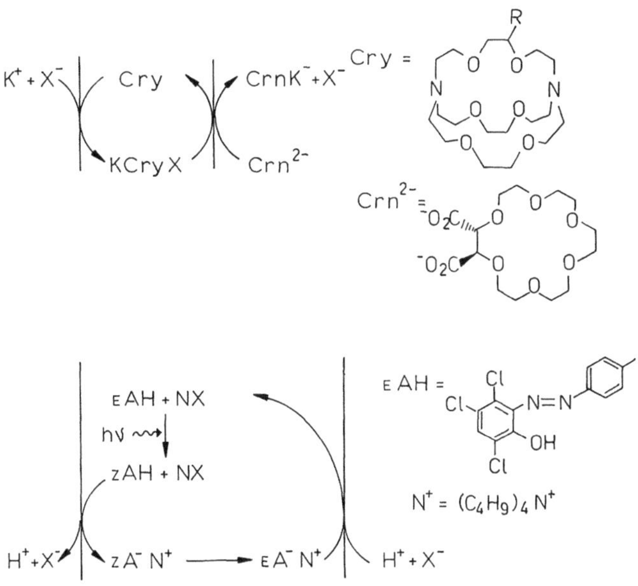

Fig. 23. Cation–anion symport reaction pumping [124, 125]

create and maintain the concentration gradient in the counterion. Although this seems to be a modest amount of energy, the energy demands of concentration gradients are small. A concentration gradient of an order of magnitude corresponds to 6 kJ of chemical potential, or 60 mV of electrical potential. In fact, ATP hydrolysis only provides 40 kJ of free energy at pH 7.4 and 37 °C yet this is used to energize the entire biochemical apparatus including creation of concentration gradients [130]. The use of complexation reactions to drive transport has been used in related systems for recovery of metals across emulsion liquid membranes [7, 8, 131, 132]. Extensions to other systems are obviously possible; this system serves to illustrate how the energetics and the kinetics of practical cycles can be related to a theoretical framework [40, 124].

The other system illustrated in Fig. 23 achieves cation–anion symport reaction pumping in a very different way. Photoisomerization of the *trans*-azo phenol carrier produces the *cis*-isomer, with a substantially augmented acidity constant [125, 126]. The *cis*-isomer has a sufficiently long lifetime to permit phase transfer reactions to occur. Thus photoisomerization effectively "pumps" HX out of the membrane into the left (light) aqueous phase. During diffusion to the opposite (dark) interface, *cis* to *trans* isomerization occurs. At the dark interface, extraction of HX occurs to re-establish the acid/conjugate base ratio appropriate for the *trans*-isomer. The net result is a pumping cycle energized by light absorption. The overall energy efficiency and the gradients which can be achieved, have not yet been analysed in detail. Even so, this is a unique system with remarkable properties. It is a formal, functional mimic of bacteriorhodopsin [16–18].

2.3 Regulation of Transport

Transport in biological systems is under tight control [1, 15]. Unlike the biomimetic systems discussed above, the natural systems operate far from equilibrium [38, 40] and are "switched on" by specific small molecules or external stimuli. Transport can then be "switched off" by removal of the activating molecule or stimulus [1, 15]. Conceptually, the activation of the transporter can be viewed as the binding of the activator to a receptor on the transporter which provokes a conformational or electronic change in the transporter [1, 2, 15]. The conformational change is supposed to enhance the affinity of the channel for its substrate or to open a gate in the channel [1, 15].

Biomimetic systems which exhibit some of the features of regulation of transport are discussed in this section. Figure 24 shows systems in which binding of a transition metal ion controls the conformation of a crown ether like moiety, thereby controlling the transport selectivity. In one case [133], binding of the $W(CO)_4$ unit forces the benzyl oxygens away from one another. The binding site can then involve only four cooperative oxygen binding sites and binding of small cations should be favoured. Experimentally, the system without tungsten bound exhibits a K^+/Na^+ selectivity of 2.3:1. Binding of $W(CO)_4$ gives a system with K^+/Na^+ selectivity of 0.5:1 [133]. This is not a reversible binding of tungsten, but two separate carrier systems. Nonetheless, the general features of conformational control of transporter selectivity are present. The other system is similar: without copper, the open chain carrier exhibits K^+:Na^+ selectivity of 2.5:1 which is enhanced to 10:1 by addition of cuprous ion to the system [134]. This would be a formal mimic of "activation" of a transport system, were it not for the fact that transport rate is depressed at the same time transport selectivity is enhanced [134].

Transporter selectivity can also be regulated by pH. The proton plays the role of the bound "activator" or stimulus which controls selectivity. One such system is illustrated in Fig. 25 [135]. The carrier is a dicarboxylic acid related to

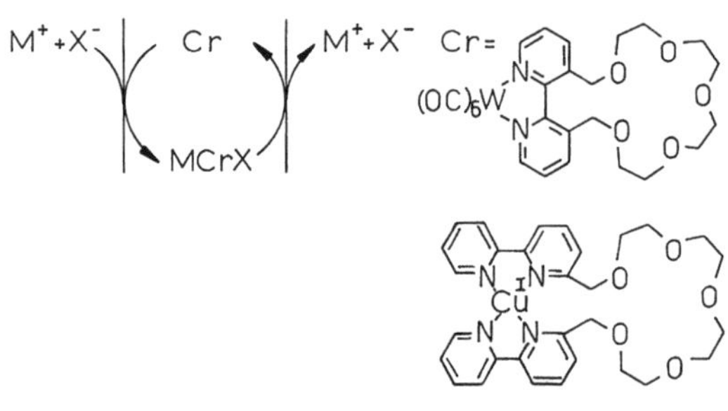

Fig. 24. Allosteric control of transport selectivity [133, 134]

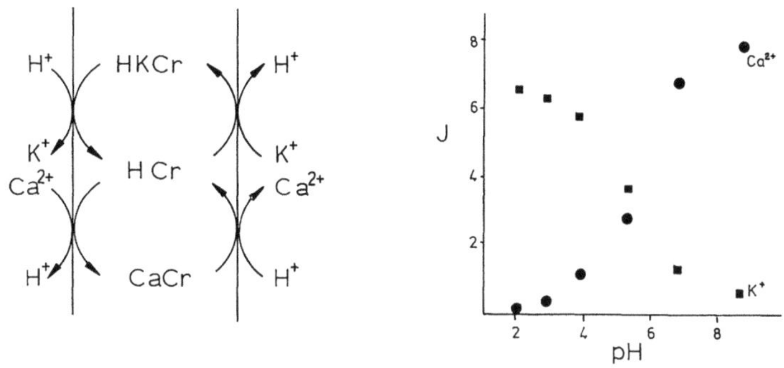

Fig. 25. pH control of transport selectivity [135]

the carrier *syn*-$(8NH)_2$ 18-$6A_2$ (Fig. 11). Double deprotonation will favour extraction of divalent ions to yield complexes such as CaCr (lower cycle Fig. 25). At lower pH, monodeprotonation will favour extraction of monovalent ions to give KHCr complexes (upper cycle Fig. 25). The balance between the two is controlled by the pH of the left aqueous phase. The selectivity $Ca^{2+}:K^+$ changes from 10:1 at pH 8.6 to 1:160 at pH 2.3 [135]. A related system involving K^+/Na^+ selectivity controlled by pH presents smaller selectivity control but is conceptually similar [136].

Redox-switching of transport can be controlled in a variety of ways. One is to act on a redox-active substrate which is, for example, reduced by external means and extracted efficiently. This is the basis of a system for the recovery of copper [131, 138]. More biomimetic are the thiol–disulfide interconversions, described by Shinkai, shown in Fig. 26; in each case the oxidized (disulfide) form of the ligand presents a more organized binding site for metal cations. As a result, oxidation produces carriers with a greater ability to extract and transport alkali metal cations by cation anion symport gradient pumping. Although transport of all ions is enhanced, the larger cavities, and bis-crown ether cavities, enhance the transport selectivity of large vs small cations [142]. The central problem in these systems is the oxidation to form polydisulfide materials. At least for the benzo crown ether at the bottom of Fig. 25, this problem has been solved, as the oxidation by iodine and by flavin derivatives proceeds via a "template" mechanism [142]. Thus transport by the reduced form can be stimulated by addition of oxidant [142]. The reverse "switching off" process proceeds very efficiently as well using external reductant [142].

Shinkai and his associates have also illustrated possibilities for photo-switching of membrane transport using the photoisomerization of azo crown ethers as illustrated in Fig. 27. The systems are of two types: systems in which the *cis*-azo form of the carrier has enhanced activity as a carrier [143–147] (light "on", switch "on") and systems in the *cis*-azo form of the carrier has depressed carrier activity [148] (light on, switch "off"). Several strategies for the former are illustrated. Of these, the provision of an anionic cap results in the most marked

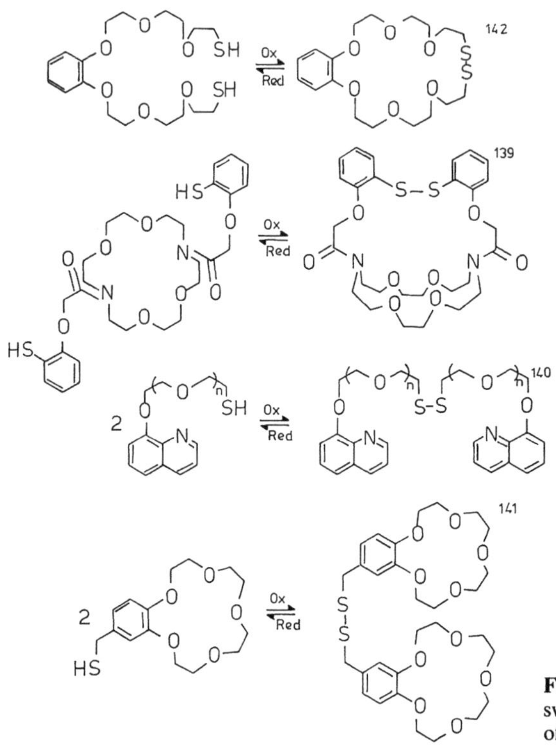

Fig. 26. Dithiol–disulfide redox switched carriers for redox switching of transport [139–142]

enhancement of flux and selectivity [145]. The composite effect of pH and light in these systems has not been investigated.

3 Polymer Membranes

The various types of polymer membranes were discussed in the introductory Sect. 1.1. The goal of this section is to survey the properties of polymer membranes which set them apart from other types of membranes with respect to biomimetic functions. Examples include relay transport and switching by thermal and electrical means.

The relay mechanism (Fig. 1) requires the ionophoric sties within the membrane to be fixed in place. This might be achieved by simple entanglement of mobile carriers in a solvent–polymer membrane [20, 149, 150]. Some diffusion of ionophore will occur, but transport of ions could occur faster by means of hopping between sites. At least under conditions of imposed electrical gradient, this has been demonstrated [151]. The better alternative, one that compels relay transport, is to bind the ionophore site to the polymer backbone. Free diffusion as a carrier is thus completely inhibited and any transport which occurs must be via a relay. Figure 28 illustrates some examples.

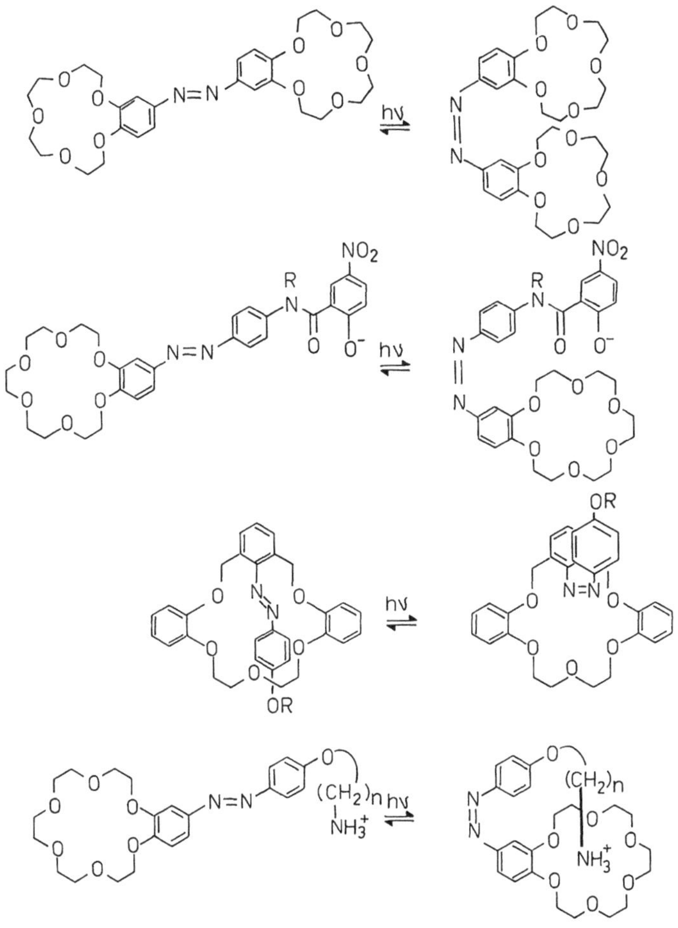

Fig. 27. Photoactive carriers for photo control of transport [143–148]

For cation transport, it is only necessary to prepare polymers of known carriers such as the crown ethers illustrated in Fig. 28. These can be pendant to the main chain [152–155] or integral to the main chain [156]. Cation proton antiport systems appear to be more active, but there are several complicating factors. In general, the mass balance of typical cation transport experiments is poor; when all species concentrations are monitored, this can be shown to be due to additional competing transport cycles [153]. These involve cation dependent anion–proton symport, in essence a relay for anions as well as a relay for cations.

Anion–proton symport transport is simpler to achieve as all that is required is an anion exchange membrane with a weakly basic site incorporated. Examples include imidazole [159] and acridine [158] derivatives but any N-

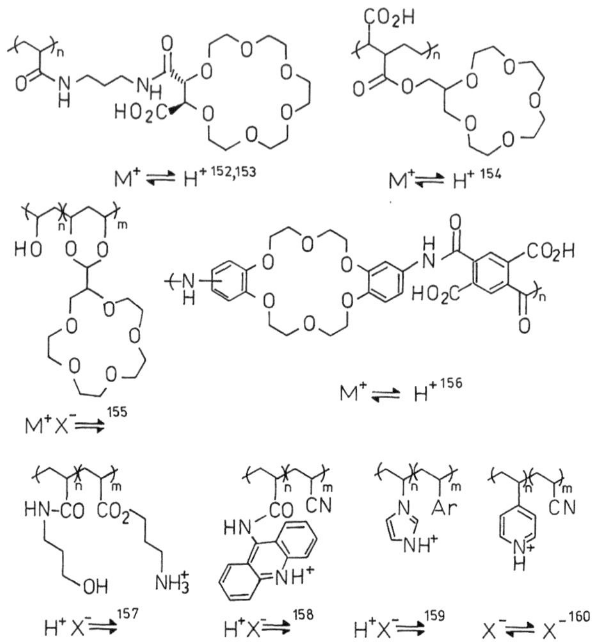

Fig. 28. Polymer membranes incorporating relays [152–158]. (XY ⇌, symport of X and Y; X ⇌ Y, antiport of X and Y)

heterocycle should behave similarly. A system based on amide/ester tautomerism provides a means to strip the anions on the basic side of the membrane [157], equivalent to the enforced release discussed previously (Fig. 13) [85, 86]. Strongly basic anion exchangers can be used for anion–anion antiport transport [160].

The relay membranes developed to date illustrate two points for further biomimetic studies. The first is that although these systems are rather easy to prepare and transport occurs readily, it is very difficult to extract mechanistic information from them. The polymers present a mixture of sites and configurations which depend on processing variables which are often loosely controlled [153]. Thus the relay created has a range of structures, none of which can be observed in isolation. The second point concerns the reversibility of most systems. In general equilibria will be established at the two interfaces resulting in a variety of different complexes at the interface. Each one of these could initiate a transfer into the membrane, thereby initiating a relay. Extractions of low probability which can be ignored in carrier systems are more important in relay systems and competing pathways develop easily [153]. Relay based membranes are thus firmly in the realm of "black box" devices.

A further insight into the relay mechanism comes from the study of polymer–liquid crystal composite membranes containing crown ether carriers (Fig. 29)

[32, 161, 162]. The basic membrane consists of a polycarbonate support, mixed with the liquid crystalline material EBBA. The EBBA undergoes a phase transition from a crystalline to a liquid crystalline phase about 305 K and undergoes further melting to an isotropic liquid about 340 K. At each phase transition, the thermal molecular motion of the EBBA and its viscosity change drastically. Thus permeation of simple solutes such as gases depends dramatically on temperature [32]. Incorporation of crown ether carriers into the system results in thermal control of ion transport due to control of diffusion of the ionophore complex. Above 305 K the systems of Fig. 29 both behave as simple cation–anion symport systems where the transport is mediated by a mobile carrier [162]. Below 305 K, the hydrocarbon crown ether system still has some transport capabilities while the fluorocarbon system does not; below 305 K, Crn(F) composite membranes are impermeable to cations. The difference lies in the aggregation of the carriers within the EBBA crystalline matrix. The hydrocarbon crown ether is apparently completely dispersed throughout the membrane phase while the fluorocarbon crown ether undergoes phase separation to microheterogeneous domains. As pictured in Fig. 29, site-to-site hopping can occur in the former case, but not in the latter due to the aggregate–aggregate separations which are large on a molecular scale [162].

An alternative type of polymer membrane with some interesting biomimetic characteristics is formed by impregnation of porous nylon capsules with bilayer forming amphiphiles [34]. The "semipermeable microcapsules" [163] or "artificial cells" [164] produced have been used to encapsulate a number of functional enzyme systems from natural sources [165]. In a biomimetic context, Okahata and coworkers have explored a variety of interesting systems based on these types of membranes [32, 166–171]. The support membrane is formed by interfacial polymerization of diamines in a water droplet floating in an organic

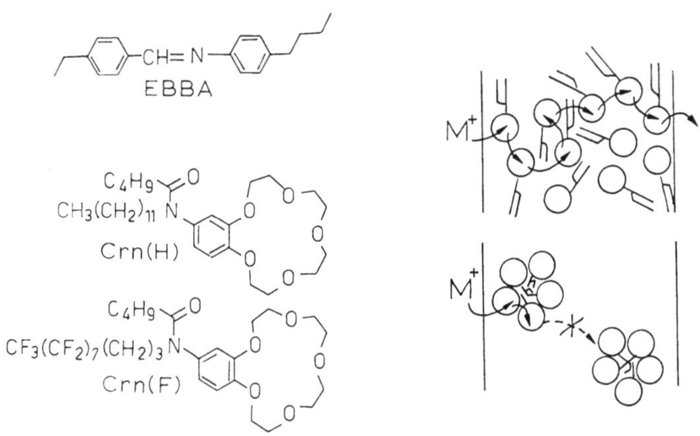

Fig. 29. Polymer–liquid crystal-carrier ternary composite membrane for thermo switched transport [162]

solvent containing di- and tri-acid chlorides. Robust porous capsules about 2 mm in diameter, with a wall thickness of 5 μm are formed [171]. Impregnation with natural and artificial bilayer forming amphiphiles gives "corked" capsule membranes (Fig. 30). The capsules are completely sealed only when the bilayer membranes are in the gel state; melting to a more fluid liquid crystalline phase results in leakage across the capsule wall [166].

More interesting are the ways in which the gel to liquid crystal phase transition can be controlled by external "signals". One example utilized synthetic anionic phospholipids held at a temperature close to the phase transition temperature for the bilayer. Initially, the capsules are sealed to NaCl, but addition of Ca^{2+} provokes NaCl efflux [167]. Addition of the Ca^{2+} complexing agent EGTA reverses the effect and reseals the capsule. Another example used carboxylate and phosphate amphiphiles as the bilayer forming component. The fluidity of the membrane is then controlled by the pH of the external solution. This system is also reversibly switched on and off by changing the pH between 7 (charged head group, gel state bilayers, permeation "off") and 2 (uncharged head group, fluid and disordered lipid layer, permeation "on") [168]. Permeation can also be controlled by an applied external electrical field [169, 170]. In the presence of the applied field, the permeation becomes independent of temperature indicating that the defects or pores formed in the bilayers are held open by the applied field. This is reminiscent of the behavior expected of a channel which should be independent of the fluid state of the bilayer [1, 2]. There is also a voltage dependence hence this is a crude mimic of a voltage gated pore [1].

Polymer liquid crystal composite membranes can be prepared with bilayer forming amphiphiles [172], providing a formal link with capsule type membranes. The key study reported that a photo switching of permeability of ions and water could be achieved [172]. The bilayer component contained a *trans*-azo linkage; photoisomerization to *cis* apparently produces defects in the bilayer

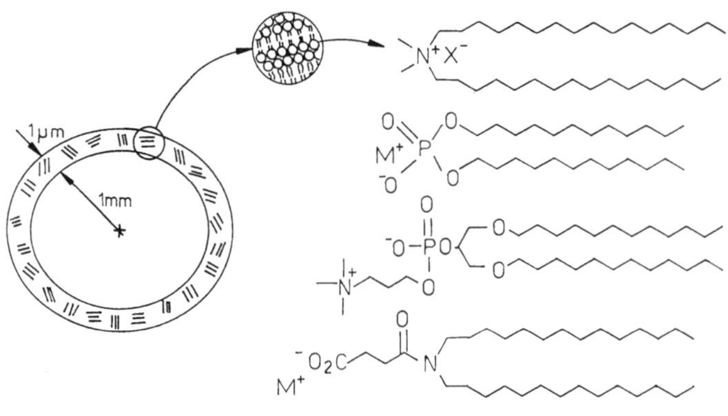

Fig. 30. Bilayer impregnated capsule membranes

for the transfer of solutes. The encouraging conclusion to be drawn from all these polymer–bilayer systems is the apparently simple mechanical and molecular mechanisms which operate. Obviously the real picture is more complex, but this simple conceptual framework serves for the simple design of new systems.

As a final note on polymer membrane systems, polymerized vesicles offer some clear biomimetic features [173, 174]. Reliable methods for the preparation and characterization of polymer coated or internally polymerized vesicle bilayers have received considerable attention. These membranes are clearly more robust and less permeable to ions and small molecules than unpolymerized systems. Unfortunately, few studies of transporters in such membranes have appeared (see Refs. [51, 173]). In view of the novel properties associated with polymer–bilayer membranes, the study of polymerized vesicles is clearly worthwhile.

4 Channels in Bilayer Membranes

The creation of artificial biomimetic pore structures in bilayer membranes poses some severe challenges from both a synthetic and an analytical perspective. Natural bilayer membranes have a thickness of about 40 Å. To create a hollow tube 40 Å long with a pore diameter of 4–6 Å requires the assembly of a structure with a molecular weight of about 3000–4000 g/mol as a minimum estimate. By the standards of directed synthesis, defined nonpolymeric targets of this molecular weight with characterized structures, present a formidable synthetic problem.

Having created a biomimetic ion channel molecule by whatever means, bilayer membrane transport studies pose a further set of problems. The common experimental vehicles of vesicles or black lipid membranes are of limited stability and present problems of reliable fabrication even by experienced workers [175]. Transport through black lipid membranes is typically followed by examination of the conductivity changes provoked by the channel former [2, 3]. For a primitive biomimetic channel these might be small and the channel might be continually "open". As a result, the already small signal associated with black lipid membrane conductivity would simply appear as noise. Transport in vesicle systems can be followed by NMR methods [59–62], by pH-stat measurement of proton efflux [87, 88] or by UV or fluorescence changes of entrapped indicator dyes [176]. All these methods depend on the volume entrapped by the vesicles. A UV method which is useful for vesicles of 5000 Å diameter may be essentially useless for 500 Å diameter vesicles due to the hundred-fold decrease in entrapped volume (and signal). Yet vesicle systems are dynamic and subject to strong perturbations in shape, size and structure by additives [177–179] (such as a biomimetic transporter as a possible example).

Despite these real experimental problems, there have been a number of approaches to the fabrication of channels in bilayer membranes.

4.1 Mimics of Antibiotics and Toxins

As with the mobile carrier mimics, naturally occurring channel and pore forming antibiotics have provided a basic model from which to elaborate biomimetic systems. There are three main types of naturally occurring channel or pore forming molecules of low (<3000 g/mol) molecular weight: i) oligopeptide toxins such as gramacidin A [2,3], ii) oligopeptide toxins such as alamethecin or mellitin [180] and, iii) polyene antibiotics such as amphotericin and nystatin [181]. From a strategic point of view these three classes are fundamentally different.

The gramacidin A channel is formed by end-to-end dimerization of two molecules, each of which adopts a Π_{DL} helical conformation with inward turning carbonyl oxygens [1–3, 13]. The helices are primarily stabilized by turn-to-turn hydrogen bonding and the interaction of hydrophobic side chains with the lipid bilayer. An ion traverses the channel down the centre of the helix and the dynamic interactions of the ion and the channel walls have been analyzed in detail [2,3]. Figure 31 shows the structure of gramacidin A and some oligopeptide mimics which apparently act as ion transporters via Π_{DL} helices [182–184]. However, other closely related DL di- and tetra-peptide oligomers which also form Π_{DL} helices apparently do not transport ions via the helix core [185]. It is thus reasonably clear that the side chains somehow mediate the ion transfer process or the channel–lipid interaction to create ion channels. This specificity limits the generality of the conclusions to be drawn about close structural and functional mimics of gramacidin. Indeed, simple substitution of four phenylalanine residues for the four tyrosine residues of gramacidin A results in an active transporter which behaves by the alternate mechanism discussed below for alamethecin [186]. From a biomimetic design perspective, it is obvious that a hollow tube with internal ion binding sites long enough to span a bilayer membrane, can act as a channel transporter. It is less obvious how to achieve this reliably from amino acid precursors.

The second class of oligopeptide toxins are represented by alamethecin, trichotoxin, melittin and various synthetic mimics (Fig. 32) [180, 187–190]. These materials form helical segments, both α helics and Π_{DL} (or β helices) which apparently aggregate to form dimer to tetramer sized associations which are ion

HOC–N–LVal–Gly–LAla–DLeu–LAla–DVal–LVal–DVal–LTrp
–DLeu–LTrp–DLeu–LTrp–DLeu–LTrp CO_2∼OH
Gramicidin A

poly (DLeu–LTrp) [182] poly (DTrp–LLeu) [182] poly (DPhe–LLeu) [182]

poly (DVal–LVal–DVal–LVal) [183] poly (DGlu(OBzl)–LGlu(OBzl)) [184]

Fig. 31. Gramicidin A and mimics which transport via Π_{DL} helices [182–184]

AcAib-Pro-Aib-Ala-Aib-Ala-Gln-Aib-Val-Aib-Gly-Leu-Aib-Pro
-Val-Aib-Aib-Glu-Gln-PheOH
Alamethecin F

AcAib-Gly-Aib-Leu-Aib-Gln-Aib-Aib-Aib-Ala-Aib-Pro-Leu-Aib
-Iva-Glu-ValOH
Trichotoxin A

HOCGly-Ile-Gly-Ala-Val-Leu-Lys-Val-Leu-Thr-Gly-Leu-Pro-Ala
-Leu-Ile-Ser-Trp-Ile-Lys-Arg-Lys-Arg-Gln-GlnNH
Mellittin

poly (DVal-LAla)[185] poly (DAla-LVal-DVal-LAla)[185]

Boc(Ala-Aib-Ala-Aib-Ala)$_4$OCH$_3$[187]

Boc(Aib-Ala)$_5$-Gly-Ala-Aib-Pro-Ala-Aib-Aib-Glu-GlnOCH$_3$[187]

poly (Ala)$_n$ $n=15,20,25$[188] poly (Ala-Ala-Gly)$_n$ $n=4,5$[189]

Leu-Leu-Leu-Ala-Leu-Leu-Gln-Leu-Leu-Phe-Gly-Leu-Leu-Ala
Leu-Leu-Leu-Glu[190]

Fig. 32. Oligopeptide toxins and mimics which transport via α helix aggregates [185–190]

conducting. The conductivity behavior of this class of oligopeptides is voltage dependent [187]. Analysis of synthetic mimics suggests that all that is required is a lipophilic helical segment of appropriate length. The inactive pore is formed by protein/lipid phase separation in which the helix dipoles stabilize one another in antiparallel arrays. The pore is turned "on" by a transmembrane potential which acts on the helix dipoles to open voids between the helices [180, 187].

In a similar mode of action, the polyene antibiotics, amphotericin and nystatin act to form aqueous pores in bilayer membranes by aggregation of 10–15 molecules in each plane of the bilayer. The aggregates then "dimerize" end-to-end to complete the pore [1, 181]. The aggregation is presumably driven by the amphiphilic nature of the antibiotic: the polyene edge interacts more favorably with lipid and other polyene edges while the polyhydroxyl edge interacts more favorably with water and other polyhydroxyl edges [181]. The pores created are of much "looser" structure than the gramicidin channel and undoubtedly have a distribution of sizes and activities. For the purposes of creating a mimic, the key features of the amphotericin pore are aggregation or phase separation in the plane of the bilayer and amphiphilic edges of the aggregating species to provide the water filled channel. This gives rather loose control of pore morphology.

Direct structural mimics of amphotericin have not been reported, but a conceptually similar pore by aggregation has been examined by Furhop [191–195]. The membrane in this system is a monolayer vesicle formed from macrocyclic tetraester derivatives bearing two polar head groups, known as bolaform amphiphiles or bolaamphiphiles [191, 194]. The lipid core thickness is about 15–20 Å, which considerably reduces the size requirements of a pore former. The pyromellitate ester of monensin (Fig. 33), in an extended conformation is approximately the same length as the membrane forming bolaamphile and renders the vesicles permeable to Li^+ [192]. The Li^+ efflux can be partly blocked by bis-quaternary ammonium salts of the appropriate length [195].

Examination of CPK models of monensin pyromellitate does not suggest any particular favorable interactions which could lead to oligomers. At the concentrations used, the pore former constitutes about 10 wt% of the total

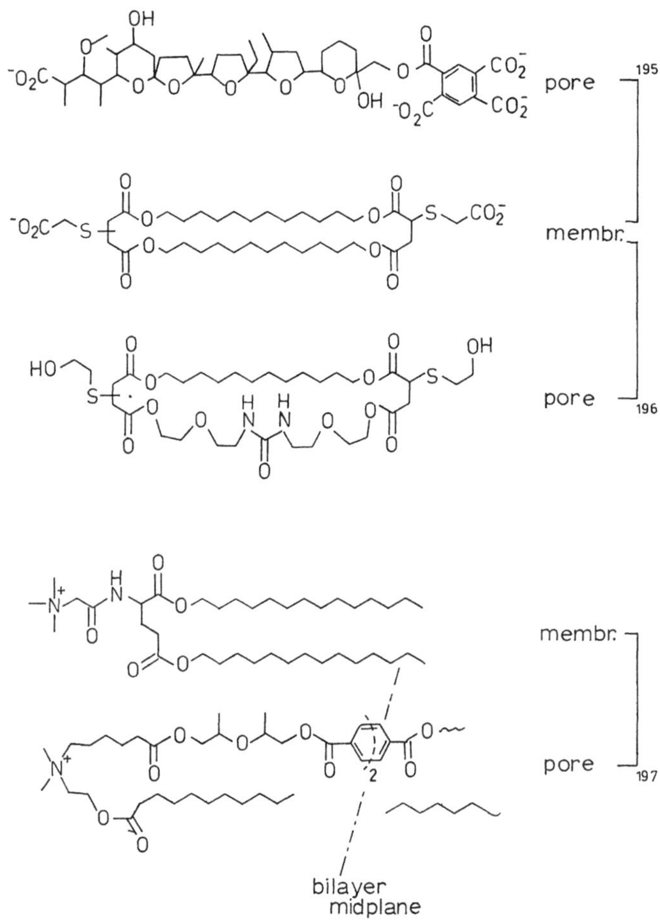

Fig. 33. Bolaform Amphiphile Pore Formers which mimic Amphotericin [195–197]

"lipid" present [194], thus it is possible that simple phase separation occurs to yield defect zones at the phase boundaries or within the monensin enriched regions.

We have recently made more explicit amphotericin mimics within this system [196] as illustrated in Fig. 33. The membrane contained a bis-fluorescein bolaamphile (10 wt%) as a pH indicator to monitor the collapse of an imposed pH gradient. The vesicles are quite "leaky" and the base rate for gradient collapse is significant ($t_{1/2} \sim 100$ sec). Nonetheless, the pore former mimics shown in Fig. 33 can increase the rate by up to a factor of five at concentrations of 0.25 wt%. The head group charge must be zero; both cationic (NH_3^+) and anionic ($-SO_3^-$, CO_2^-) groups on the same structures are inactive. The apparent kinetic order in pore former is 2 or 3 [196]. All these observations are consistent with the formation of small aggregates which increase ionic permeability.

Another type of amphotericin mimic has been described by Kunitake [197] (Fig. 33 bottom). The membrane in this case is a bilayer vesicle formed from a glutamate diester with entrapped riboflavin. Permeation of alkali is followed by fluorescence quenching. These vesicles are also leaky, but addition of either amphotericin/cholesterol (1:1, 0.12 wt%) or the pore formers shown in Fig. 33 accelerates the fluorescence quenching reaction by 2 to 5 fold [197]. The transport is clearly a function of membrane fluidity consistent with the formation of permeable aggregates.

4.2 Synthetic Channel Forming Compounds

Even though the goal of biomimetic ion transport studies is to reflect on natural ion transporters, we need not be held to any slavish imitation. At the same time we might reasonably explore without the guidance of an original to imitate. This is nowhere more true than when confronting natural transporter proteins. In favorable cases, structures associated with transport functions can be observed by electron microscopy. These are apparently fluted tubes 50–60 Å in diameter which narrow to a single constriction near the bilayer midplane [198]. The structural groups which create the constriction are unknown. Alternatively, the aggregate pores formed in bacterial outer membranes (porin channels) appear to provide 10 Å aqueous holes at the centre of several subunit contacts [199]. Even when the primary and secondary structures are fully or partly known, the nature of the "channel" which undoubtedly exists is hardly apparent [18].

We are, therefore, free to consider virtually any "hollow tube" structures as potential synthetic channels.

One possibility is found in the solid state structure of a tartaric acid crown ether derivative [200]. The crown ethers are held in face-to-face stacks at about 5 Å separation. Potassium ions in the structure occupy two types of sites: a typical crown ether site at the centre of the macrocycle and site midpoint between crown ethers. The latter site represents an ion in mid hop in a relay [200]. Translation of this crystal architecture into a single molecule of defined

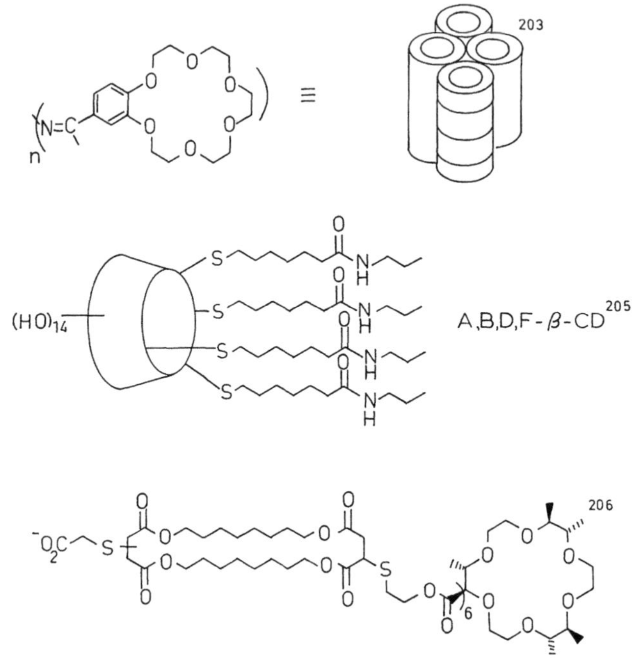

Fig. 34. Synthetic Ion Channels [201–207]

structure has involved herculean synthetic effort, but to date has not produced channel sized structures [201–202] (Fig. 34).

Face-to-face crown ethers have been created by polymerization of isocyanide substituted benzo 18-crown-6 [203–204] (Fig. 34). The polymers are stacks of crown ethers arranged in four columns about a central poly(isocyanide) core and from molecular weight measurements are about 40 Å long. The polymers will transport Co^{2+} across bilayer vesicle walls albeit at a very slow rate ($t_{1/2}$ ~ 30 min) [204]. The transport is, however, independent of the fluid state of the bilayer and is completely consistent with a channel constructed of several relay type binding sites.

Another artificial channel forming compound has been reported by Tabushi [206], based on a cyclodextrin framework (Fig. 34). The cyclodextrin provides a hollow tube inlet structure. In some ways this structure is a relative of gramacidin in that end-to-end dimers are required to span a bilayer. The cyclodextrin half channel mediates Cu^{2+} transport across egg lecithin vesicle bilayers ($t_{1/2}$ ~ 5 min) with an apparent kinetic order of 2. This half channel also transports Co^{2+} but at a slower rate and with an apparent first order dependence [205].

The final channel illustrated in Fig. 34 uses a crown ether hexaacid as the controlling framework [206]. The parent hexaacid, and other crown ethers derived from tartaric acid, exhibit a strongly marked tendency to place carboxylate derived groups in axial positions [207]. The macrocyclic tetraester arms, derived from the related systems in Fig. 33, are thus expected to project above and below the crown ether plane. Polar head groups, and the rigid spacer units in the arms are expected to orient the channel in the bilayer with the crown ether providing a restriction at the bilayer midplane. In the event, this unimolecular channel mediates collapse of an imposed pH gradient across egg phosphatidyl choline derived vesicles [206]. The collapse is independent of added proton uncouplers indicating that cation–proton antiport is achieved. Specific accelerations of up to a hundred-fold over the base leakage of the membrane have been achieved. The relative activity of the channel is comparable on a molar basis to melittin at 0.01 wt%, and vesicle permeability can be induced at sub nanomolar concentrations of added channel [206]. Exploration of this system with side discriminated arms as in Fig. 33 is ongoing.

The central conclusion for this section must be to emphasize that simple structures can be manipulated in simple ways to achieve effective ion transport across vesicle membranes. This stands in contrast to enzyme mimics. The active sites of natural enzymes are comparatively well understood when compared with transport proteins. Even so, artificial enzyme mimics fall short of natural enzymes by many orders of magnitude in activity [208]. Even in the virtual absence of structural guidance from natural transporters, biomimetic ion transport systems can be constructed to have specific activities directly comparable to natural systems [197, 206]. Although experimentally more complex, ion transport systems are clearly simpler to construct.

5 References

1. Finean JB, Coleman R, Michell RH (1984) Membranes and their cellular functions. Blackwell, Oxford
2. Lauger P (1985) Angew. Chem. Int'l. Ed'n. Engl. 24: 905
3. Pullmann A (1988) Pure Appl. Chem. 60: 259
4. Oxford English Dictionary (1971) Oxford University Press, Oxford
5. Li NN (1978) J. Membr. Sci. 3: 265
6. Cussler EL, Evans DF (1980) J. Membr. Sci. 6: 113
7. Izatt RM, Deardon DV, McBride DW, Oscarson JL, Lamb JD, Christensen JJ (1983) Sep. Sci. Tech. 18: 1113
8. Izatt RM, Clark GA, Bradshaw JS, Lamb JD, Christensen JJ (1986) Sep. Purif. Meth. 15: 21
9. Izatt RM, LindH GC, Bruening RL, Bradshaw JS, Lamb JD, Christensen JJ (1986) Pure Appl. Chem. 58: 1453
10. Shinkai S, Manabe O (1984) Top. Curr. Chem. 121: 67
11. McBride DW, Izatt RM, Lamb JD, Christensen JJ (1984) In: Atwood JL, Davies JED, MacNichol J (eds) Inclusion Compounds, Academic Press, London, vol. 3, p 571
12. Painter GR, Pressman BC (1982) Top. Curr. Chem. 101: 83
13. Urry DN (1985) Top. Curr. Chem. 128: 175
14. Cammann K (1985) Top. Curr. Chem. 128: 219
15. Hofer M (1981) Transport across biological membranes. Pitman, London
16. See Curr. Top Membr. Transpt. Vol 21 (1984)
17. Houslay MD, Stanley KK (1982) Dynamics of Biological Membranes, John Wiley, New York
18. Ovchnnikov YA (1987) Chemica Scripta 27B: 149
19. Racker E (1987) Chemica Scripta 27B: 131

20. Morf WE (1981) Principles of ion-selective electrodes and membrane transport. Elsevier, Amsterdam
21. Fendler JM (1982) Membrane mimetic chemistry. John Wiley, New York
22. Wong KH, Yagi K, Smid J (1974) J. Membr Biol 18: 379
23. Fyles TM, Malik-Diemer VA, Whitfield DM (1981) Can. J. Chem. 59: 1734
24. Rosano HL, Schulman JH, Weisbach JB (1961) Ann. N.Y. Acad. Sci. 92: 457
25. Lamb JD, Izatt RM, Garrick DG, Bradshaw JS, Christensen JJ (1981) J. Membr. Sci. 9: 83
26. Pressman BC, de Guzman NT (1975) Ann. N.Y. Acad. Sci. 264: 373
27. Tsukube H (1983) J.C.S. Perkin Trans. I 1983: 29
28. Danesi PR (1984) Sep. Sci. Tech. 19: 857
29. Danesi PR (1984) J. Membr. Sci. 20: 231
30. Cahn RP, Li NN (1974) Sep. Sci. Tech. 9: 505
31. Lee KH, Evans DF, Cussler EL (1978) A.I. Ch. E. Journal 24: 80
32. Kajiyama T, Washiza S, Kumano A, Terada I, Takagamagi M, Shinkai S (1985) J. Appl. Poly. Sci. Appl. Poly. Symp. 41: 327
33. Shinkai S, Nakamura S, Tahiki S, Manabe O, Kajiyama T (1985) J. Am. Chem. Soc. 107: 3363
34. Okahata Y, Enna G, Tagichi K, Seki T (1965) J. Am. Chem. Soc. 107: 5300
35. Lakshmninarayanaiah N (1969) Transport Phenomena in Membranes, Academic, New York
36. Cussler EL (1976) Multi Component Diffusion, Elsevier Amsterdam
37. Szabo G, Eisenman G, Laprade R, Ciani SM, Krasne S (1973) In: Eisenman G (ed) Lipid Bilayers and Antibiotics, Dekker, New York, vol. 2, p 181
38. Katchalsky A, Curran DF (1967) Nonequilibrium thermodynamics in biophysics, Harvard University Press, Cambridge
39. Zwolinski BJ, Eyring H, Reese CE (1949) J. Phys. Colloid Chem. 53: 1426
40. Goddard JD (1985) J. Phys. Chem. 89: 1825
41. Reusch CF, Cussler EL (1975) A.I. Ch. E. Journal 19: 736
42. Caracciolo F, Cussler EL, Evans DF (1975) A.I. Ch. E. Journal 21: 160
43. Lamb JD, Christensen JJ, Izatt SR, Bedke K, Astin MS, Izatt RM (1980) J. Am. Chem. Soc. 102: 3399
44. Lamb JD, Christensen JJ, Oscarson JL, Nielsen BL, Asay BW, Izatt RM (1980) J. Am. Chem. Soc. 102: 6820
45. Yoshida S, Hayano S (1986) J. Am. Chem. Soc. 108: 3903
46. Sakim M, Hayashita T, Yamabe T, Igawa M (1987) Bull. Chem. Soc. Jpn. 60: 1289
47. Kirch M (1980) Transport of alkali metal cations by cryptates. Thesis, Université Louis Pasteur, Strasbourg
48. Behr JP, Kirch M, Lehn JM (1985) J. Am. Chem. Soc. 107: 241
49. Fyles TM (1985) J. Membr. Sci. 24: 229
50. Fyles TM (1987) Can. J. Chem. 65: 884
51. Ovchinnikov YA, Ivanov VY, Skorb AM (1974) Membrane Active Complexones, Elsevier, Amsterdam
52. Westley JN (ed) (1982) Polyether Antibiotics, Marcel Dekker, New York
53. Dobler M (1981) Ionophores and their structures. J. Wiley and Sons, New York
54. Taylor, RW, Kaufmann RF and Pfeiffer DR (1982) In: Westley JN (ed) Polyether Antibiotics, Marcel Dekker, New York, vol 1, p 103
55. Ashton R, Steinrauff LK (1970) J. Mol. Biol. 49: 547
56. Choy EM, Evans DF, Cussler EL (1974) J. Am. Chem. Soc. 96: 7085
57. Bolte J, Demuynck C, Jeminet G, Juillard J, Tissier C (1982) Can. J. Chem. 60: 981
58. Yang W, Yamauchi A, Kimizuka H (1987) J. Membr. Sci. 31: 109
59. Ting DZ, Hagan PS, Chan SI, Doll JD, Springer CS (1981) Biophys. J. 34: 189
60. Riddel FG, Arumugam S, Brophy PJ, Cox BG, Payne MCH, Southon TE (1988) J. Am. Chem. Soc. 110: 734
61. Riddel FG, Arumugam S, Cox BG (1987) J.C.S. Chem. Comm. 1987: 1890
62. Bartsch RA, Grandjean J, Laszlo P (1983) Biochem. Biophys. Res. Comm. 117: 340
63. Tsukube H, Takagi K, Higashiyama T, Iwachido T, Hayama N (1986) J.C.S. Chem. Comm. 1986: 448
64. Suzuki K, Tohda K, Sasakura H, Inoue H, Tatsuta K, Shirai T (1987) J.C.S. Chem. Comm. 1987: 932
65. Samat A, Bibout MEM, Elguero J (1985) J.C.S. Perkin I 1985: 1717
66. Delser CM, Young MEM, Tom-Kun J (1980) Biochem. 19: 6194

67. Kimura S, Imanishi Y (1985) in Kiso Y (ed) Peptide Chemistry 1985, Protein Research Foundation, Osaka
68. Lifson S, Felder CE, Shanzer A, Libman J (1987) In: Izatt RM, Christensen JJ (eds) Progress in Macrocycle Chemistry, vol 3, John Wiley, New York, p 241
69. Gardner JO, Beard CC (1978) J. Med. Chem. 21: 357
70. Wierenga W, Evans BR, Woltersom JA (1979) J. Am. Chem. Soc. 101: 1334
71. Kuboniwa H, Yamaguchi K, Hirao A, Nakahana S, Yamazaki N (1982) Chem. Lett. 1982: 1937
72. Kuboniwa H, Nagami S, Yamaguchi K, Hirao A, Nakahana S, Yamazaki N (1985) J.C.S. Chem. Comm. 1985: 1468
73. Hiratani K, Taguchi K, Sugihara H, Iio K (1984) Bull. Chem. Soc. Jpn. 57: 1976
74. Umen MJ, Scarpa A (1970) J. Med. Chem. 21: 505
75. Hamilton JA, Steinrauf LK, Braden B (1975) Biochem. Biophys. Res. Comm. 64: 151
76. Kauffman RF, Taylor RW, Pfieffer DR (1980) J. Biol. Chem. 255: 2735
77. Fyles TM, Malik-Diemer VA, McGavin CA, Whitfield DM (1982) Can. J. Chem. 60: 2259
78. Dulyea LM, Fyles TM, Whitfield DM (1984) Can. J. Chem. 62: 498
79. Fyles TM, Whitfield DM (1984) Can. J. Chem. 62: 507
80. Strezelbicki J, Bartsch RA (1982) J. Membr. Sci. 10: 35
81. Charewicz WA, Bartsch RA (1983) J. Membr. Sci. 12: 323
82. Bartsch RA, Charewicz WA, Kang SI J. Membr. Sci. 17: 97
83. Fyles TM (1985) J.C.S. Farad I 82: 617
84. Inokuma S, Yabusa K, Kuwanmura (1984) Chem. Lett. 1984: 607
85. Shinkai S, Kinda H, Sone T, Manabe O (1982) J.C.S. Chem. Comm. 1982: 125
86. Shinkai S, Kinda H, Arargi Y, Manabe O (1983) Bull. Chem. Soc. Jpn. 56: 559
87. Thomas C, Sauterey C, Castaing M, Gary-Bobo CM, Lehn JM, Plumere P (1983) Biochem. Biophys. Res. Comm. 116: 981
88. Castaing M, Morel F, Lehn JM (1986) J. Membr Biol. 89: 251
89. Behr JP, Lehn JM (1973) J. Am. Chem. Soc. 95: 6108
90. Bacon E, Jung L, Lehn JM (1980) J. Chem. Res. (S) 1980: 136
91. Bacon E, Jung L, Lehn JM (1980) J. Med. Chem. 15: 89
92. Sugiura M, Yamaguchi T (1984) Sep. Sci. Tech 19: 623
93. Tsukube H (1983) J. Poly. Sci. Poly. Lett. Ed'n 21: 639
94. Tsukube H (1983) J.C.S. Chem. Comm. 1983: 970
95. Tsukube H, Takagi K, Higashiyama T, Iwachido T, Hayama N (1985) J.C.S. Perkin Trans. II 1985: 1541
96. Katyaoka H, Katagi T (1987) Tetrahedron 40: 4519
97. Newcomb M, Toner JL, Helgeson RC, Cram DJ (1979) J. Am. Chem. Soc. 101: 4941
98. Sollner K, Shean GM (1964) J. Am. Chem. Soc. 86: 1901
99. Shean GM, Sollner K (1966) Ann. N.Y. Acad. Sci. 137: 758
100. Tsukube H (1982) Angew Chem. Int'l Ed'n Engl. 21: 304
101. Maruyama K, Tsukube H, Araki T (1982) J. Am. Chem. Soc. 104: 5197
102. Tsukube H (1982) J.C.S. Perkin Trans I 1982: 2359
103. Tsukube H (1983) Bull. Chem. Soc. Jpn. 56: 1883
104. Tsukube H, Takagi K, Higashiyama T, Iwachido T, Hayama N (1986) Bull. Chem. Soc. Jpn. 59: 2021
105. Lehn JM (1983) In: Spach G (ed) Physical chemistry of transmembrane ion motions, Elsevier, Amsterdam, p 181
106. Tabushi I, Imuta J, Seko N, Kobuke Y (1978) J. Am. Chem. Soc. 100: 6287
107. Tsukube H (1983) Tetrahedron Lett. 1983: 1519
108. Tsukube H (1985) J.C.S. Perkin Trans. I 1985: 615
109. Dietrich B, Fyles TM, Hosseini MW, Lehn JM, Kaye KC (1988) J.C.S. Chem. Comm. 1988: 691
110. Hosseini MW, Lehn JM (1986) Helv. Chim. Acta 69: 587
111. Lehn JM (1985) Science 227: 1262
112. Park CH, Simmons HE (1968) J. Am. Chem. Soc. 90: 2431
113. Lehn JM, Moradpur A, Behr JP (1975) J. Am. Chem. Soc. 97: 2532
114. Shinbo T, Nishimura K, Yamaguchi T, Sugiura M (1986) J.C.S. Chem. Comm. 1986: 349
115. Anzai J, Kobayashi Y, Ueno A, Osa T (1984) Makromol. Chem. Rapid Comm. 5: 715
116. Diederich F, Dick K (1984) J. Am. Chem. Soc. 106: 8024

117. Harada A, Takahashi S (1987) J.C.S. Chem. Comm. 1987: 527
118. Anderson SS, Lyle LG, Paterson R (1976) Nature 259: 147
119. Grimaldi JJ, Boileau S, Lehn JM (1977) Nature 265: 229
120. Grimaldi JJ, Lehn JM (1979) J. Am. Chem. Soc. 101: 1333
121. Saji T. Kinoshita I (1986) J.C.S. Chem. Comm. 1986: 716
122. Shinbo T, Sugiura M, Kamo N, Kobatake Y (1901) J. Membr. Sci. 9: 1
123. Ohki A, Takagi M, Ueno K (1980) Chem. Lett. 1980: 1591
124. Fyles TM, Hansen SP (1988) Can. J. Chem. 66: 1445
125. Haberfield P (1987) J. Am. Chem. Soc. 109: 6178
126. Haberfield P (1987) J. Am. Chem. Soc. 109: 6177
127. Okahara M, Nakatsuji Y (1985) Top. Curr. Chem. 127: 37
128. Shinbo T, Kurihara K, Kobatake Y, Kano N (1977) Nature 270: 277
129. Maros Y, Nakashima T (1983) J. Phys. Chem. 87: 794
130. Mahler HR, Cordes EH (1971) Biological Chemistry, 2nd edn, Harper and Row, New York, p 36
131. Christensen JJ, Christensen SP, Biehl MP, Lowe SA, Lamb JD, Izatt RM (1983) Sep. Sci. Tech. 18: 363
132. Izatt, RM, Bruening RL, Clark GA, Lamb JD, Christensen JJ (1986) J. Membr. Sci. 28: 77
133. Rebek J. Wattley RV (1980) J. Am. Chem. Soc. 102: 4853
134. Nabeshima T, Inaba T, Furukawa (1987) Tetrahedron Lett. 28: 6211
135. Hriciga A, Lehn JM (1983) Proc. Nat'l. Acad. Sci. USA 80: 6426
136. Dugas H, Brunet P, Desroches (1986) Tetrahedron Lett. 27: 7
137. Ohki A, Takeda T, Takagi M, Ueno K (1982) Chem. Lett. 1982: 1529
138. Ohko A, Takeda T, Takagi M, Ueno K (1983) J. Membr. Sci. 15: 231
139. Shinkai S, Inuzuka K, Manabe O (1983) Chem. Lett. 1983: 747
140. Shinkai S, Inuzuka K, Hara K, Sone T, Manabe O (1984) Bull. Chem. Soc. Jpn. 57: 2150
141. Shinkai S, Minami T, Araragi Y, Manabe O (1985) J.C.S. Perkin Trans II 1985: 503
142. Shinkai S, Inuzuka K, Miyazaki O, Manabe O (1985) J. Am. Chem. Soc. 107: 3950
143. Shinkai S, Nakaji T, Ogawa T, Shigematsu K, Manabe O (1981) J. Am. Chem. Soc. 103: 111
144. Shinkai S, Shigematsu K, Sato M, Manabe O (1982) J.C.S. Perkin Trans I 1982: 2735
145. Shinkai S, Minami T, Kusano Y, Manabe O (1982) J. Am. Chem. Soc. 104: 1967
146. Shinkai S, Miyazuki K, Manabe O (1985) Angew Chem. Int'l Ed'n Engl. 24: 866
147. Shinkai S, Miyazaki K, Manabe O (1987) J.C.S. Perkin Trans I 1987: 449
148. Shinkai S, Yoshida T, Miyazaki K, Manabe O (1987) Bull. Chem. Soc. Jpn. 60: 1819
149. Sugiura M (1987) J. Colloid Interfac. Sci. 81: 385
150. Thoma AP, Viviani-Nauer A, Arvantis S, Morf WE, Simon W (1977) Anal. Chem. 49: 1567
151. Wipf HK, Olivier A, Simon W (1970) 53: 1605
152. Fyles TM, McGavin CA, Thompson DE (1982) J.C.S. Chem. Comm. 1982: 924
153. Dulyea LM, Fyles TM, Robertson GD (1987) J. Membr. Sci. 87: 8521
154. Kimura K, Sakamoto H, Yoshinaga M, Shono T (1983) J.C.S. Chem. Comm. 1983: 978
155. Kimura K, Yoshinaga M, Kitazawa S, Shonto T (1983) J. Poly. Sci. Poly. Chem. Edn. 21: 2777
156. Sakamoto H, Kimora K, Shono T (1986) Eur. Poly. J. 22: 97
157. Ogata N, Sanui K, Fujimura H (1981) J. Appl. Poly. Sci. 26: 4149
158. Yoshikawa M, Ogata H, Sanui K, Ogata N (1983) Poly. J. 15: 609
159. Yoshikawa M, Imashiro Y, Sanui K, Ogata N (1984) J. Membr. Sci. 20: 189
160. Yoshikawa M, Imashiro Y, Yatsuzuka Y, Sanuik, Ogata N (1985) J. Membr. Sci. 23: 347
161. Shinkai S, Torigoe K, Manabe O, Kajiyama T (1986) J.C.S. Chem. Comm. 1986: 933
162. Shinkai S, Torigoe K, Manabe O, Kajiyama T (1987) J. Am. Chem. Soc. 109: 4458
163. Chang TMS, MacIntosh FC, Mason SG (1966) Can. J. Physiol. Pharmacol. 44: 115
164. Chang TMS (1972) Artificial cells, Charles C Thomas, Springfield
165. Rosenthal AM, Change TMS (1980) J. Membr. Sci. 6: 329
166. Okahata Y, Lim HJ, Nakumura G, Hachiya S (1983) J. Am. Chem. Soc. 105: 4855
167. Okahata Y, Lim HJ (1984) J. Am. Chem. Soc. 106: 4696
168. Okahata Y, Seki T (1984) J. Am. Chem. Soc. 106: 8065
169. Okahata Y, Hachiya S, Seki T (1984) J.C.S. Chem. Comm. 1984: 1377
170. Okahata Y, Hachiya S, Ariga K, Seki T (1986) J. Am. Chem. Soc. 108: 2863
171. Okahata Y (1986) Acc. Chem. Res. 19: 57
172. Kumano A, Niwa O, Kajiyama T, Takayanage M, Kunitake T, Vano K (1984) Poly. J. 16: 461
173. Fendler JH (1984) Acc. Chem. Res. 17: 3

174. Fendler JH (1984) Science 223: 888
175. Szoka F Papahadjopoulos (1978) Proc. Nat'l. Acad. Sci. USA 75: 4174
176. Moore HP, Raftery M (1980) Proc. Nat'l Acad. Sci. USA 77: 4509
177. Mitchell DJ, Ninham BW (1981) J.C.S. Farad. Trans. II 77: 601
178. Ninham BW, Evans DF (1986) Farad. Diss. Chem. Soc. 81: 1
179. Evans DF (1988) Langmuir 4: 3
180. Bernheimer AW, Rudy B (1986) Biochem. Biophys. Acta 864: 123
181. Finkelstein A, Holz R (1973) In: Eisenman G (ed) Membranes: Lipid bilayers and antibiotics, Marcel Dekker, New York, vol 2, p 357
182. Heitz F, Lotz B, Colonna-Cesari F, Spack G (1980) In: Srinivasan R (ed) Biomolecular structure, conformation, function and evolution, Pergamon, Oxford, vol 2, p 59
183. Heitz, F, Lotz B, Détriché G, Spach G (1980) Biochem. Biophys. Acta 596: 137
184. Spach G (1978) Chimia 32: 124
185. Heitz F, Détriché G, Trudelle F, Spach G (1981) Macromolecules 14: 57
186. Spach G, Heitz F, Trudelle Y (1983) In: Spach G (ed) Physical chemistry of transmembrane ion motions, Elsevier, Amsterdam, p 375
187. Jung G, Katz E, Schmitt H, Voges KP, Menestrina, Boheim G (1983) In: Spach G (ed) Physical chemistry of transmembrane ion motions, Elsevier, Amsterdam, p 349
188. Heitz F, Spach G, Seta P, Gavach C (1982) Biochem. Biophys. Res. Comm. 107: 481
189. Goodall MC (1973) Anch. Biochem. Biophys 157: 514
190. Spach G, Merle Y, Molle G (1985) J. Chimie Phys 82: 719
191. Furhop JH, Ellerman K, David HH, Mathieu (1982) Angew Chem. Int'l Ed'n Engl. 21: 440
192. Furhop JH, Liman U (1984) J. Am. Chem. Soc. 106: 4643
193. Furhop JH, Liman U, David HH (1985) Angew, Chem. Int'l Ed'n Engl. 24: 339
194. Furhop JH, David HH, Mathieu J, Liman U, Winter MJ, Boekema E (1986) J. Am. Chem. Soc. 108: 1785
195. Furhop JH, Fritsch D (1986) Acc. Chem. Res. 19: 130
196. Fyles TM, James TD, Zojaji M (1988) unpublished observations
197. Kunitake T (1986) Ann. NY Acad. Sci. 471: 70
198. Barr PH, Gage PN (1984) Curr. Top. Membr. Transpt. 21: 35
199. Benz R (1984) Curr. Top Membr. Transpt. 21: 199
200. Behr JP, Lehn JM, Dock AC, Mores D (1982) Nature 295: 526
201. Heng R (1985) Polyfunctional Crown Ethers. Thesis, University Louis Pasteur, Strasbourg
202. Behr JP, Bergdoll M, Chevrier B, Dumas P, Lehn JM, Moras D (1987) Tetrahedron Lett 28: 1989
203. Van Beijnen AJM, Nolte RJM, Zwicker JN (1982) Rec. Trav. Chim. Pays. Bas. 101: 409
204. Neevel JC, Nolte RJM (1984) Tetrahedron Lett. 25: 2263
205. Tabushi I, Kuroda Y, Yokota K (1982) Tetrahedron Lett. 23: 4601
206. Carmichael VE, Dutton PJ, Fyles TM, James TD, Swan JA, Zojaji M (1989) J. Am. Chem. Soc. 111: 769
207. Dutton PJ, Fyles TM, McDermid SJ (1988) Can. J. Chem. 66: 1097
208. R. Breslow (1986) Ann. NY Acad. Sci. 471: ix

Lariat Ethers in Membranes and as Membranes

George W. Gokel and Luis Echegoyen
Department of Chemistry, University of Miami, Coral Gables, Florida 33124, U.S.A.

Lariat ethers are macrocyclic (crown ethers) that have one or more sidearms appended to the ring. The sidearms usually contain Lewis basic donor groups when cation binding properties are desired. By combining both a macroring and a sidearm, the advantages of both three-dimensional encapsulation and flexibility are realized. Flexibility is important because cation transport requires rapid complexation and release, especially the latter if it is to be effective. The structural, cation binding, and transport properties are discussed herein. It is also possible to prepare macrocycles having lipophilic sidearms that do not contain any donor group. If lipophilic is enough, these lariat ethers self-assemble into micelles, niosomes, or vesicles. If appropriately designed, these compounds may assemble into smaller, more readily characterized structures. Examples of all of these cases are illustrated herein.

Bioorganic Chemistry Frontiers, Vol. 1

1 Classes of Cation-Binding Molecules

Numerous compounds that have cation binding properties are now known. Their range can be gleaned from any of the monographs dealing with polyether antibiotics [1], complexones [2], or ionophores [3]. Although a detailed discussion of the range of structural types and properties exhibited by cation binders is well beyond the scope of this chapter, several properties are common. These include, multiple oxygen binding sites, a ring or the ability to dimerize into a ring-like structure, and the ability to encapsulate the cation. All of these properties are clear in the remarkable cation binder known as valinomycin [4].

1.1 Valinomycin

Valinomycin is a cyclododecadepsipeptide. In other words, it is a cyclic molecule containing 12 amino acid and hydroxy acid residues. Its total ring size is 36 atoms. Its structure is given by the simplified formula:

$$-[NH-CH(i\text{-}Pr)-CO-O-CHMe-CO-NH-CH(i\text{-}Pr)-CO-O-CH(i\text{-}Pr)-CO]_3-$$

This four element subunit is repeated three times to give a total of twelve acid elements attached to nine isopropyl groups and three methyl groups. The molecule is thus quite lipophilic. It is interesting that the amide and ester carbonyl groups alternate. It is more interesting that the stereochemistry of the chiral subunits fit the pattern (L, L, D, D)$_3$. The presence of so many D-amino acids and this stereochemical pattern seem remarkable at first, but this arrangement is a simple solution to a complicated problem.

Although it appears that valinomycin is far too large to effectively encircle K^+, it is selective for this cation over several other biologically important cations such as Li^+, Na^+, and Ca^{2+}. Indeed, the selectivity ratios are as follows for these cations (in methanol): K^+/Na^+, 6,000 and K^+/Ca^{2+}, 80 [4]. This selectivity is extraordinary, especially when compared to even the most K^+-selective cryptand, [2.2.2] [5]. This remarkable selectivity is accomplished by a folding of the molecule into a tennis-ball-seam [6] arrangement. As a result, the two-dimensional macrocycle folds into a three-dimensional array of donor groups. This three-dimensional arrangement is maintained by six hydrogen bonds to the amide carbonyl groups [4]. This leaves only the six ester carbonyl residues available for binding. Fortunately, these ester carbonyl groups form an almost perfect octahedron about the cavity available to a cation. This structural arrangement also has the consequence of using the less polar carbonyl groups to bind the less charge dense K^+ cation rather than the smaller Na^+ or Ca^{2+} ion.

A further and no less remarkable aspect of valinomycin's structure is the high percentage of D-amino and hydroxy acid subunits. Stryer [7] has noted that D-amino acids "are never found in proteins". Of course, D-amino acids are common in peptides produced by bacteria but the observation is no less remarkable. What purpose do these D-amino acids have? When bends occur in a peptide derivative, there is usually a proline present at that position. Proline is unique among the essential amino acids because it is cyclic and contains a secondary nitrogen atom. The problem with proline is that it is a rigid molecular structure, a fact that can readily be confirmed by an examination of CPK, space-filling molecular models. That D-acids have the opposite chirality of the L-isomers, requires that they have the opposite *geometry*. A tennis-ball-seam arrangement requires repeating bends and these are provided by the alternating chirality without imposing rigidity on the structure. The latter is important to any successful ionophore because not only must a cation be bound and transported, it must be released after transport if the ionophore is to be useful.

1.2 Binding Strength and Dynamics

The reaction between a cation complexing agent and a cation may be characterized in several ways. The reaction itself may be represented as ligand + cation = complex. The extent of this reaction is characterized by an equilibrium constant, usually called K_S, which in turn depends upon solvent, temperature, etc. It is a simple, but occasionally overlooked, fact that K_S depends on both the *rates* of complexation and decomplexation: $K_S = k_{complex}/k_{decomplex}$, $k_{bind}/k_{release}$, or k_1/k_{-1}. The latter rate is of special importance in transport since it, along with diffusion, determines the efficacy of the agent in question. We compare the rates of complexation and decomplexation in homogeneous solution for valinomycin, 18-crown-6, and [2.2.2]-cryptand below in Table 1.

Table 1. Kinetic data for several K^+-selective ligands

Ligand	$k_{complex}$	$k_{release}$	K_S	Refs.
Valinomycin[a]	4.0×10^7	1.3×10^3	3.1×10^4	[4]
18-crown-6[b]	4.3×10^8	3.7×10^6	115	[8]
[2.2.2]-cryptand[b]	7.5×10^6	38	2.0×10^5	[9]
[2.2.2]-cryptand[a]	4.7×10^8	1.8×10^{-2}	2.6×10^{10}	[5]

[a] Value determined in anhydrous methanol;
[b] value determined in water

In all of the cases shown in Table 1, complexation occurs at a significant rate. Release of the cation by 18-crown-6 is so fast that the overall binding constant is low. The cryptand exhibits excellent binding strength in both water and methanol, but the release rate is extremely slow. An advantage of the cryptands is that they form encapsulating complexes whereas the more dynamic crown ethers are essentially two-dimensional binders. The beauty of the valinomycin structure is thus apparent. It is a three-dimensional binder that can completely envelop and solvate a cation. It remains flexible by using a combination of chirality and hydrogen bonds to maintain its binding conformation. It was, in large measure, to mimic these remarkable properties, that we developed the compounds known as the lariat ethers.

2 Lariat Ethers

The syntheses of valinomycin [10] and analogs of it [11] are daunting tasks although both have been undertaken. Much has been learned about these structures but the complexity of these compounds limits the utility of this approach if the intent is to develop novel ionophores and/or to extend existing properties. Our analysis of valinomycin suggested that three-dimensionality and flexibility were both required for the development of a novel ionophore. The cryptands offered three-dimensionality but lacked the dynamics that were apparently required. The crown ethers were flexible and dynamic, but lacked the three-dimensionality exhibited by valinomycin. We therefore undertook the synthesis of a family of molecules having both a macroring and a sidearm.

Our original concept of lariat ethers [12] involved a ring and a sidearm that cooperated in binding by virtue of sidearm donor groups. This is not actually the case for many of the compounds discussed in this article. Even so, the sidearm may play other important roles and we thus refer to these compounds as lariat ethers although they do not fit our original definition and concept.

2.1 Carbon-Pivot Lariat Ethers

18-Crown-6 is a landmark compound in the development of the crown ether area [13]. Its remarkable K^+ selectivity and high binding strength [8, 14] are both notable. Our plan was to use the 3n-crown-n framework as the basis for complexation strength and selectivity in the lariat ethers. To the macrocyclic ring we would append a sidearm containing one or more donor atoms. These donors would, at least at first, be neutral, Lewis basic donors, rather than anionic residues. This was done because cooperation between ring and sidearm was desired and the presence of a charge might dominate binding and selectivity and substantially alter flexibility.

It is possible to attach a sidearm either to the 2-position of a 3n-crown-n molecule or to a macroring oxygen atom. In the latter case, an unstable oxonium salt would result so this option was not pursued. The sidearm could be attached to the macroring carbon atom with another carbon or with a hetero-atom. Again, the latter option would lead to an acetal [15] if the heteroatom was oxygen so carbon was selected instead. From these considerations, the subunit comprising the point of attachment emerges as glycerol, $HOCH_2CHOHCH_2OH$. The sidearm may be attached to one of the primary hydroxyl groups and the remaining primary and secondary hydroxyls form one macroring subunit. In fact, the syntheses were accomplished either from epi-chlorohydrin or allyl chloride. The glycidyl or allyl ether was prepared and then either hydrolyzed [12] or oxidized [16] to give a $ROCH_2CHOHCH_2OH$ unit. The two hydroxyl groups could then be incorporated into a 3n-crown-n macrocycle in the normal fashion [16]. A typical synthesis is shown in Scheme 1.

A variety of carbon-pivot structures were studied in our early work. These compounds were mostly 15-crown-5 derivatives [16]. This family of compounds

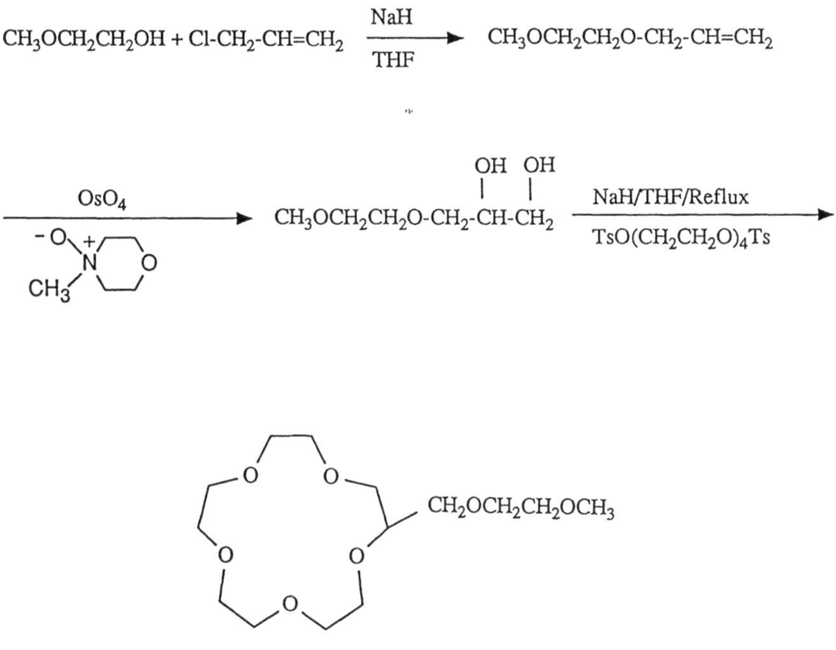

Scheme 1.

has been expanded considerably by the efforts of Okahara and his coworkers who have prepared both 15- and 18-membered ring, carbon-pivot structures having a methyl or other alkyl group at the pivot carbon [17]. Our expectation of ring-sidearm cooperation was realized, evidence for which was obtained from extraction constant measurements [12]. Thus, the two compounds shown below as **1** and **2** extracted 15.7% and 6.4% of the available sodium picrate from water into dichloromethane solution. This compares with 7.6% extracted by 15-crown-5. Clearly, the position of the methoxy group is crucial to the efficacy of these compounds. Moreover, the cyclization yield observed for **1** is about 70%, nearly double that obtained for **2**.

Although these two-phase results were extremely encouraging in terms of the lariat ether concept, they could not be confirmed by (homogeneous) K_S meas-

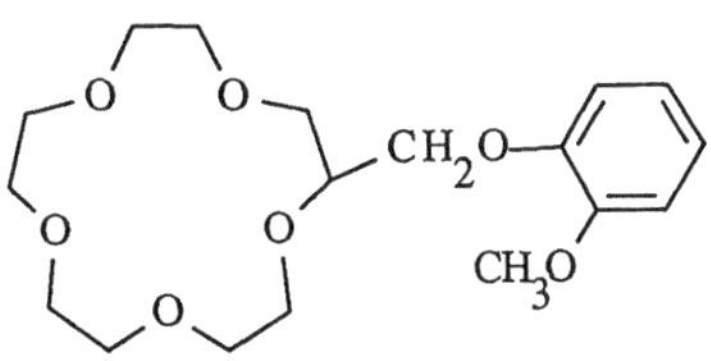

1

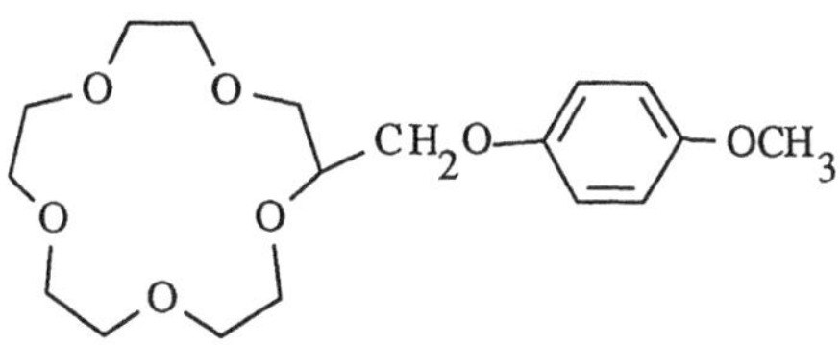

2

urements conducted in methanol solution [16]. The $\log_{10} K_S$ values (in 90% v/v aqueous methanol) obtained for 15-crown-5, **1**, and **2** were, respectively, 2.97, 2.97, and 2.56. The corresponding values for 15-crown-5 and **1** in anhydrous methanol were 3.27 and 3.24. Thus, the sidearm certainly did enhance cation binding when the donor group was appropriately placed, but cation binding strength, at least as judged by homogeneous equilibrium constant measurements, was not significantly enhanced over the parent macrocycle. We attributed part of this problem to the inherent "sidedness" of carbon-pivot lariat ethers. Since carbon is non-invertable, the sidearm can interact with a ring-bound cation from only one side of the macroring. We thus undertook the synthesis of compounds we believed would be more flexible and, consequently, more versatile.

2.2 Nitrogen-Pivot Lariat Ethers

When nitrogen replaces a macroring oxygen atom in a crown ether, it becomes possible to place the sidearm directly on the heteroatom. Since nitrogen undergoes rapid and facile inversion, the problem of sidedness is eliminated. Nitrogen is sensitive to oxidation so the nitrogen-pivot lariat ethers are slightly less stable than the carbon-pivot structures, but this has proved not to be a significant problem in our studies. When this effort was in the planning stage, studies with CPK molecular models suggested that for 18-membered ring systems, a three-carbon bridge placed the sidearm donor directly over the macroring-bound cation. We decided to use a two-, rather than, three-carbon bridge for several reasons. First, the conformations of the two-carbon sidearm should more closely resemble those of the macroring and, because of the interest in crown ethers, more is known about the ethyleneoxy subunit than about the propyleneoxy residue. Synthetic access was also more obvious in the two-carbon series for multiple-donor sidearm compounds. It should also be noted that use of a diethanolamine precursor permitted preparation of both 15- and 18-membered ring systems quite readily whereas the latter were less accessible in the carbon-pivot series.

2.3 Synthetic Access to Nitrogen-Pivot Lariat ethers

An important problem in our study of carbon-pivot lariat ether chemistry was synthetic access. It is not that such compounds are inherently difficult to prepare, although they often present a more substantial challenge than might be imagined. The problem is that the ring is formed from a substituted ethylene glycol derivative of the form R–CHOH–CH_2OH and thus require a $(CH_2CH_2O)_4$ unit for the synthesis of a 15-membered ring and a $(CH_2CH_2O)_5$ unit for the 18-membered ring systems. The former is readily accessible in the form of commercially available tetraethylene glycol. Although pentaethylene glycol has been available from several sources over the years, it has often proved impure and always quite expensive. Tetraethylene glycol is less expensive, but even the best samples contain as much as 10% of $HO(CH_2CH_2O)_nH$ impurities. These can be removed by careful (spinning band) distillation beofore incorporation into the ring. If these other oligomers are not removed, the analogous impurities that result from cyclization of the crude material are all but impossible to excise.

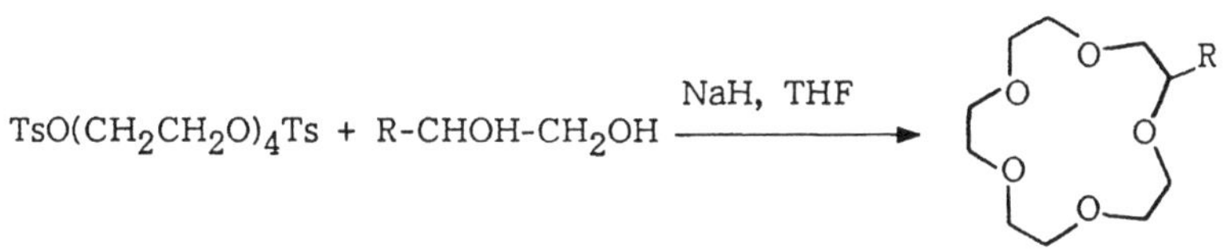

Since pentaethylene glycol is more expensive and/or more difficult to prepare and obtain, and because the boiling point is higher, the distillation is more tedious than for tetraethylene glycol. Synthesis of the 18-membered ring compounds is more difficult than the 15-membered ring structures and it is for this reason that a far larger number of all-aliphatic 15-crown-5 derivatives are known than are members of the 18-crown-6 family. In either case, synthesis is accomplished essentially as follows [16].

The synthesis of an 18-membered ring, nitrogen-pivot lariat ether has the advantage that the precursor already contains two ethyleneoxy units. Synthesis is usually accomplished by treatment of diethanolamine with the appropriate sidearm precursor in the presence of Na_2CO_3.

$$HOCH_2CH_2NCH_2CH_2OH + R\text{–}X \longrightarrow (HOCH_2CH_2)_2N\text{–}R$$

Cyclization is then effected essentially as above using a base (typically NaH) in an aprotic solvent (typically THF or DMF) in concert with a polyethylene glycol ditosylate. Fifteen-membered rings result from the reaction with triethylene glycol ditosylate (or dimesylate) and 18-membered rings are obtained from tetraethylene glycol ditosylate.

Cyclization yields are usually good for the aza-18-crown-6 compounds, often exceeding 50% [18]. The unsubstituted aza-3n-crown-n systems are accessible either by synthesis of the *N*-benzyl compounds followed by hydrogenolysis, or by the direct method of Okahara and coworkers [19]. *N*-Alkylation then affords derivatives inaccessible by the diethanolamine alkylation route.

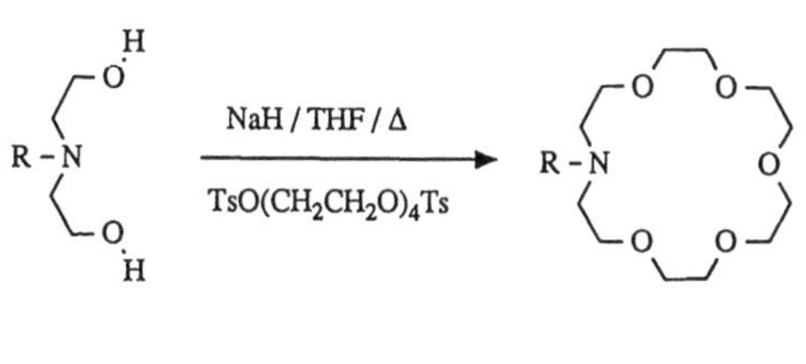

Two-armed (bibracchial) lariat ethers may be prepared by two methods as well. A single-step approach in which a primary, aliphatic amine is allowed to react with triethylene glycol diiodide, affords a variety of symmetrically substituted, *N,N'*-4,13-diaza-18-crown-6 derivatives [20]. This approach offers the advantage of simplicity and the disadvantage that the products often prove difficult to purify. A more traditional, two-step approach has also been developed [21]. This method requires more steps, but yields are usually higher and the approach is more versatile. Thus, the two-step approach may be used to prepare 15-membered ring BiBLE derivative, whereas the single step approach, shown below, affords only 18-membered rings. A number of other syntheses have been reported [22].

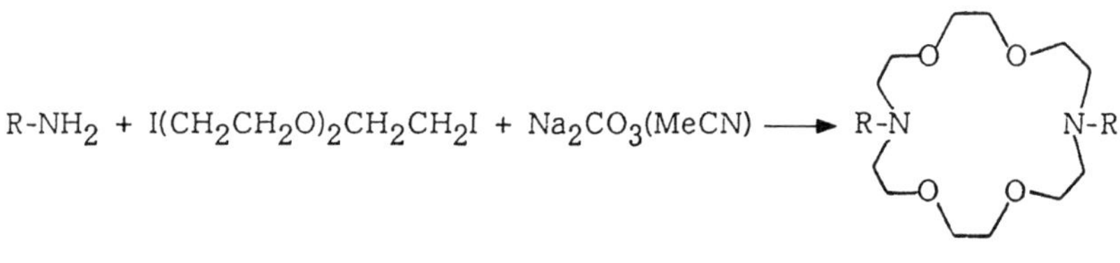

A final note on syntheses is in order. Many of the compounds discussed in this chapter are glycine derivatives. These compounds are generally obtained by the reaction of the sidearm as an alkylating agent with an aza-3n-crown-n. The alkylating agent is prepared by treatment of the appropriate alcohol with commercially-available chloroacetyl chloride [23]. The latter reaction occurs readily and usually in high yield. The sequence is shown below for the synthesis of a highly lipophilic ester derivative [18].

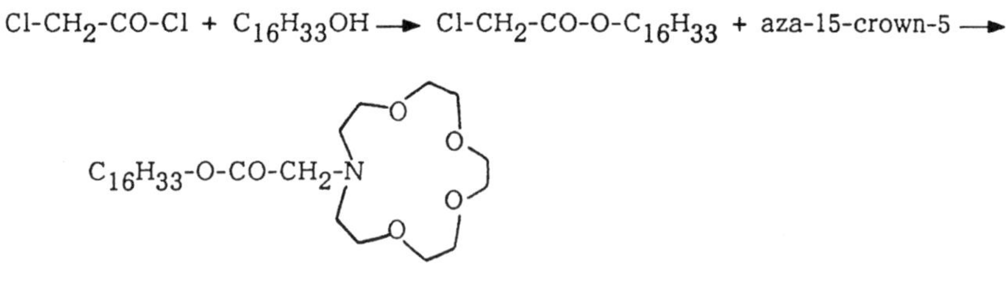

2.4 Cation Binding by Lariat Ethers

A brief comparison of cation binding by nitrogen- and carbon-pivot lariat ethers will suggest the remarkable differences between them. Cation binding constants determined in anhydrous methanol solution at $25 \pm 0.1\,°C$ using ion selective electrode methods are shown in Table 2.

Some general observations concerning the cation binding of these compounds can be made from the information in this table. First, binding by the *N*-pivot lariat ethers almost always is superior to that of the *C*-pivot structure

Table 2. Cation binding by lariat ethers

Pivot Point	Ring Size	Sidearm	Binding: Log K_S Na^+	K^+
C	15	H (15-crown-5)	3.27	3.60
C	15	$CH_2OCH_2CH_2OCH_3$	3.01	3.20
C	15	$CH_2O(CH_2CH_2O)_2CH_3$	3.13	3.50
C	15	$CH_2OC_6H_4$-2-OCH_3	3.24	3.47
C	15	$CH_2OC_6H_4$-4-OCH_3	2.90	3.18
C	18	H (18-crown-6)	4.35	6.08
N	12	$CH_2CH_2OCH_3$	3.25	2.73
N	12	$(CH_2CH_2O)_2CH_3$	3.60	ND
N	12	$(CH_2CH_2O)_3CH_3$	3.64	3.85
N	15	$CH_2CH_2OCH_3$	3.88	3.95
N	15	$(CH_2CH_2O)_2CH_3$	4.54	4.68
N	15	$(CH_2CH_2O)_3CH_3$	4.32	4.91
N	15	$(CH_2CH_2O)_4CH_3$	4.15	5.28
N	18	$CH_2CH_2OCH_3$	4.58	5.67
N	18	$(CH_2CH_2O)_2CH_3$	4.33	6.07
N	18	$(CH_2CH_2O)_3CH_3$	4.28	5.81
N	18	$(CH_2CH_2O)_4CH_3$	4.27	5.86

Note: ND means not determined

corresponding most closely to it. Second, Na^+ and K^+ binding by the 15- and 18-membered ring systems generally exceeds that of the 12-membered ring systems, although not nearly so much as might be expected from the differences in binding known for the parent macrocycles [24]. Third, K^+ binding strengths are almost always greater than Na^+ binding strengths, at least for the series of compounds presented here. This binding information will prove of interest when we consider electrochemically-switched systems, below.

2.5 Sidearm Lipophilicity in *N*-Pivot Lariat Ethers

Because cation binding strength and transport have been a focus of crown ether chemistry since Pedersen's initial report [13] of these remarkable compounds in 1967, we studied how cation binding strengths and selectivities were affected by changes in lipophilicity in single-armed [16] and two-armed lariat ether compounds [21b]. A broader range of two-armed compounds were prepared than single-armed compounds for reasons that will be described below. We slightly altered the nomenclature for these systems since the presence of two arms needed to be specified. We used the Latin word *bracchium* meaning "arm" and called the two-armed systems "bibracchial lariat ethers" or BiBLEs, for short [20]. Three-armed systems could then be readily accommodated as TriBLEs, [25] and so on. Homogeneous cation binding constants for several one- and two-armed lariat ethers are shown in Table 3.

It is apparent from the data shown in Table 3 that lipophilicity affects the cation binding strength in homogeneous solution relatively little. Some small

Table 3. Cation binding by lipophilic, nitrogen-pivot lariat ethers

Ring Size	No. of Arms	Sidearm on nitrogen	Binding strength (log K_S)[a] Na^+	K^+
15	1	*n*-butyl	3.02	2.90
15	1	*t*-butyl	2.15	2.41
15	1	Benzyl	2.77	2.61
15	1	CH_2COO-Et	4.10	4.03
15	1	CH_2COO-*t*-butyl	4.20	4.06
15	1	CH_2COO-*n*-hexyl	4.10	3.97
15	1	CH_2COO-*n*-dodecyl	4.07	3.95
15	1	CH_2COO-*n*-hexadecyl	4.11	3.99
15	1	CH_2COO-cholesteryl	4.10	4.03
15	1	CH_2COO-cholestanyl	4.12	4.03
15	1	COO-cholesteryl	< 1.5	< 1.5
15[b]	2	CH_2COO-Et	5.34	4.65
15[b]	2	Benzyl	3.59	3.13
18[c]	1	*n*-propyl	3.50	4.92
18[c]	1	CH_2COO-Et	4.67	5.92
18[c]	1	Benzyl	3.41	4.88
18[c]	2	*n*-propyl	2.86	3.77
18[c]	2	*n*-butyl	2.84	3.82
18[c]	2	*n*-hexyl	2.89	3.78
18[c]	2	*n*-nonyl	2.95	3.70
18[c]	2	*n*-dodecyl	2.99	3.80

differences are noted such as that between *n*-butyl and *t*-butyl, but these differences have more to do with solvent effects (or solvent disordering effects) [26, 27] than with lipophilicity. The glycine derivatives make an especially interesting series. First, the expected increase in binding is observed when either the ring size is increased or the number of sidearms is doubled. The increase in both cases is due to the increase in donor atom number. In the 15-membered ring series, however, the change in cation binding strength observed as the sidearm lipophilicity increases from two atoms (ethyl) to 27 atoms (cholesteryl) is negligible.

Special attention should be paid to the 15-membered ring compound having a COO-cholesteryl sidearm [23]. This compound is quite different in binding strength from the others. This is because the linkage between the steroid and the macroring is a urethane (carbamate). The lone pair electrons present on nitrogen are involved in resonance with the carbonyl group and are therefore unavailable for interaction with a ring-bound cation. This effectively reduces the number of donor groups to four for this compound. When only four donor groups are present in *N*-benzylaza-12-crown-4, the Na^+ binding affinity ($\log K_S$) is 2.08. This compares with < 1.5 as shown in Table 3. We were fortunate enough to obtain a solid state structure of this compound and both the chemical structure and the ORTEP plot are shown below in Fig. 1.

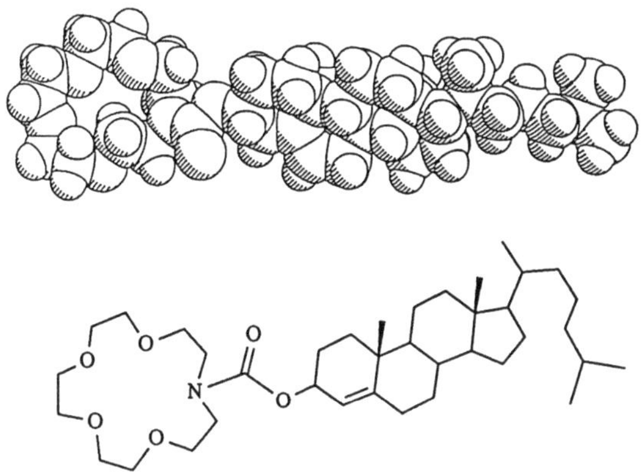

Fig. 1. Structure of a cholesteryl lariat ether compound

2.6 Dynamics of Lariat Ether Systems

By our definition, those systems having rapid binding and release *rates* are dynamic. Such compounds are also usually quite flexible, relatively open structures rather than those forming a closed, three-dimensional cavity as is the case with cryptands. The lariat ethers are typically flexible cation binders. This is apparent from the recent rate studies conducted by Eyring, Petrucci, and their coworkers [29].

We showed the binding rates for valinomycin, 18-crown-6, and [2.2.2]-cryptand in Table 1. 18-Crown-6 is in the "flexible" and "dynamic" category and [2.2.2]-cryptand is clearly more rigid. Valinomycin is remarkable because it is intermediate both in the sense of three-dimensionality and in binding dynamics. The lariat ethers likewise fall into this "intermediate" category. Using ultrasonic relaxation techniques, Eyring and Petrucci found that aza-15-crown-5, when substituted on nitrogen by a 2-(2-(2-methoxy)ethoxy)ethoxy sidearm complexed Na^+ in two steps. They determined the rate and equilibrium constants for these two steps as follows: $k_1 = 9.0 \times 10^{10}\ M^{-1}s^{-1}$, $k_{-1} = 2.1 \times 10^8\ s^{-1}$, $K_1 = 429\ M^{-1}$; $k_2 = 1.2 \times 10^7\ s^{-1}$, $k_{-2} = 1.5 \times 10^5\ s^{-1}$, $K_2 = 80$. Note that the product of these equilibrium constants is [(429)(80) =] 34,320 ($\log_{10} = 4.535$). The reported [16] binding constant for this compound with Na^+ is 4.54 (shown in Table 2).

Note that both the rates and binding constant compare favorably with those reported for valinomycin and shown in Table 1. Thus, as envisioned, the lariat ethers are dynamic and reasonably strong cation binders. Unfortunately, their cation selectivities do not really rival those of valinomycin, but some progress has been made in this direction as well.

3 Electrochemical Switching of Lariat Ether Compounds

One of the problems inherent to transport across a membrane is that at least three distinct phases are involved. Cation binding at the entry (or source) phase should be rapid and strong. Inside the lipophilic membrane, only strength is important since the bound cation has no need to exchange with the medium. At the exit (receiving) phase, cation binding should be, ideally at least, dynamic and weak. The effect of diffusion [30] should not be overlooked, but this is membrane and solvent dependent. In considering the requirements for transport as noted above, the common ligand systems present an enigma. What sort of system would be both strongly and weakly binding as required for cation uptake and release at different stages of the transport process? In an effort to develop a system that could be managed, we prepared lariat ethers which could be reversibly switched to afford strong binding at the source phase and weak binding at the receiving phase.

3.1 Reversible Reduction of Nitrobenzene-Substituted Lariat Ethers

Nitrobenzene is well-known to undergo reduction to give a radical anion [31]. This radical anion is considerably more electron rich than the parent aromatic compound. This property proved useful in the design of switchable lariat ethers. Thus nitrobenzene was incorporated into the sidearm of a carbon-pivot lariat ether [32]. It was expected, based upon a study of CPK molecular models and our previous experience with lariat ether compounds, that an *ortho*-nitrophenyl sidearm would cooperate in binding a ring-bound cation whereas the *para*-isomer would not. Both should be reducible so the extent of cooperation could be assessed by a direct comparison of these two structures. Further comparisons with neutral and/or non-reducible lariat ethers would permit us to gauge their relative binding ability prior to reduction and after it, as well.

A feeling for the inherent cation binding ability of the nitroaromatic lariat ethers can be obtained by comparing three derivatives of aza-15-crown-5. They are *N*-(2-methoxybenzyl)-, *N*-(2-nitrobenzyl)- (**3**) and *N*-(4-nitrobenzyl)aza-15-crown-5 (**4**). Their Na^+ binding constants determined in anhydrous methanol at

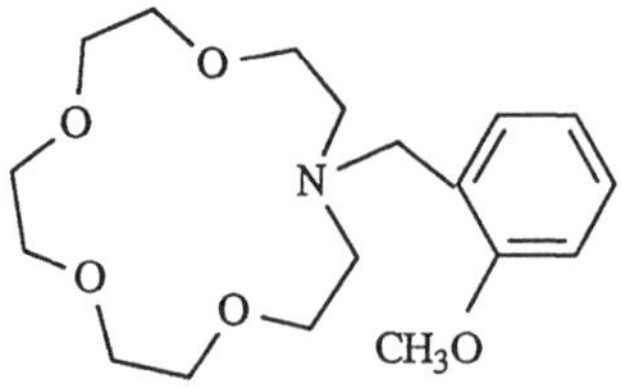

2-methoxybenzylaza-15-crown-5

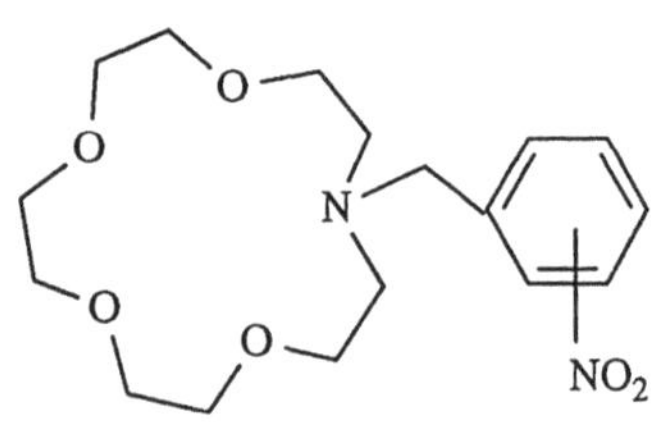

3. 2-nitrobenzylaza-15-crown-5

4. 4-nitrobenzylaza-15-crown-5

25 °C are, respectively, 3.54, 2.40, and 2.30 [30]. For comparison, consider that *N*-(2-methoxyphenyl)aza-15-crown-5 has a sodium cation binding constant of 3.86 while the *para*-isomer has log $K_S = 2.12$ [16]. The nitro ($-NO_2$) group is clearly a poor donor. Even so, we felt that when reduced, cation binding strength would be enhanced and that the *para*-isomer would be inferior in this respect to the *ortho*-isomer as seen for the neutral systems.

The cyclic voltammogram (in CH_3CN, tetrabutylammonium perchlorate) obtained for **3** showed a single, quasi-reversible redox couple typical of nitrobenzene. When 0.25 equivalent of $NaClO_4$ were added, a new redox couple was observed at more positive potential. When 0.5 equivalent of salt were present, the two couples were nearly identical in size. The (original) more negative peak diminished in intensity as the amount of salt approached 1.0 equivalent, by which point, the original couple was no longer observed. Addition of [2.2.1]-cryptand, a strong but very non-dynamic binder (see data in Table 1) was added and the original couple was completely restored [32]. This is undoubtedly due to the cryptand's poor release rate since the binding constant for the reduced lariat ether, estimated from the electrochemical data, is close to that for the cryptand [33]. When the same series of experiments were conducted on the *para*-isomer, there was some indication of a new, irreversible redox couple, but nothing appeared whose meaning was readily interpretable [32].

Redox-switched cation complexation is a fundamental and, we believe, important principle. Its application in cation transport is obvious. At the source phase, the weakly binding ligand can be "switched-on" by reduction (electrochemical or otherwise). This leads to greatly enhanced cation binding. Strong binding is maintained as the cation-ligand complex passes through the membrane phase. Finally, at the exit (receiving) phase, the reduced ligand may be "switched-off" by oxidation. The original, relatively poor binding properties of the ligand are thus restored and the cation is released. The principles of this process have proved valid in subsequent work (see below) but the reduced nitroaromatic system is too water-sensitive to be useful in this context. As a result, other reducible residues were explored.

3.2 Anthraquinone-Substituted Podands and Lariat Ethers

Anthraquinone has the ability to undergo reversible reduction just as nitrobenzene does but there are two advantages to it. First, it is capable of undergoing two separate redox reactions rather than just one. Second, its radical anion is stable in aqueous solution for months so long as oxygen is absent. The latter contrasts with the behavior of nitrobenzene radical anion which is instantly protonated in water. Unfortunately, the syntheses of anthraquinone derivatives have proved a greater challenge than anticipated. Even so, these compounds have afforded interesting and useful results.

We prepared the two simple anthraquinone derivatives shown below as **5** and **6**. Compound **5** [34] is an anthraquinone podand [35] while **6** is a lariat ether. Although nitroaromatic podand compounds failed to show interesting

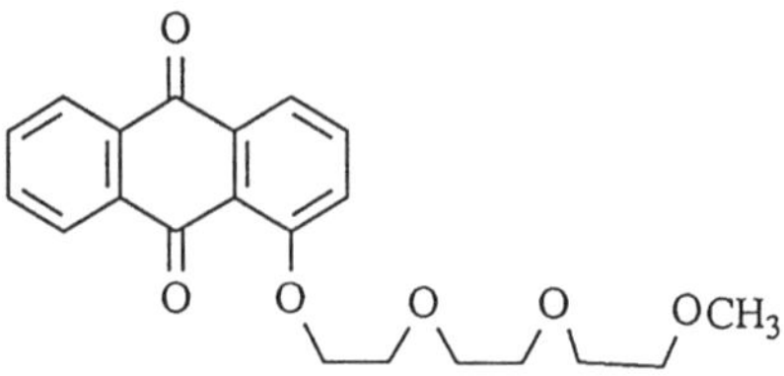

5

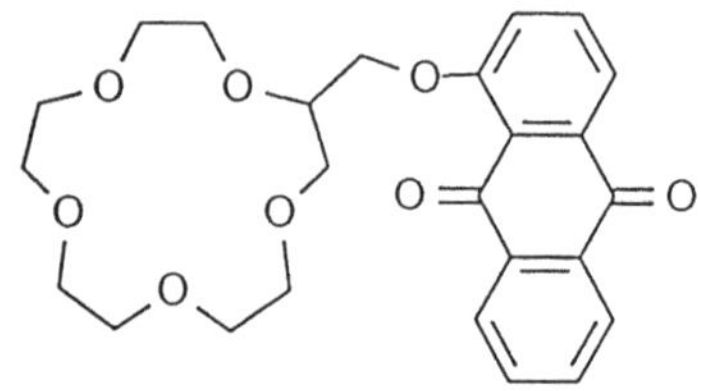

6

redox-initiated cation binding behavior [31], the anthraquinone podands have proved to be excellent ligands when reduced [34].

4 Highly Lipophilic, Redox-Switched Ionophores

In order to achieve effective membrane transport, a combination of "switchability" and lipophilicity is required. We were both pleased and somewhat surprised in our early work to discover that reduced anthraquinone podands are as effective as reduced anthraquinone-substituted lariat ethers in cation complexation [34]. 1-(2-Methoxy(2-ethoxy(2-ethoxy(2-ethoxy(2-ethoxy))))) anthraquinone is illustrated below.

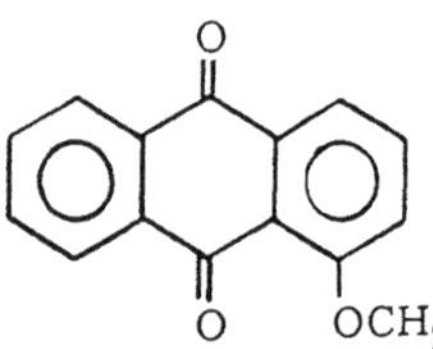

$OCH_2CH_2OCH_2CH_2OCH_2CH_2OCH_2CH_2OCH_3$

The anthraquinone-podand shown above cannot transport Li^+ across a dichloromethane bulk membrane, at least not to any appreciable extent. This is because Li^+ is a charge-dense [36] cation and requires substantial solvation to stabilize it in solution. When anthraquinone is (electrochemically) reduced to its radical anion, it is charged and capable of stabilizing the Li^+ cation. In a specific experiment, no significant transport was observed for the neutral ligand during more than two hours of observation. When reduced (potential of -1.0 V *vs* Ag wire), this ligand was able to transport Li^+ across a CH_2Cl_2 model membrane system with a rate of 2.2×10^{-7} mol/h [37]. Direct EPR evidence showed that a strong ion pair existed in solution between Li^+ and the reduced ligand.

Two coordinated approaches are underway to enhance the prospects for "switched" transport. Several types of lipophilic sidechains have been attached to the anthraquinone podand pictured above. Attachment to the 5- or 8-positions have been accomplished using ether or ester links. As expected, the hydrolytic stability of the esters is inferior to that of the ethers, although they are synthetically more accessible. Indeed, two soon to be published studies [38] will show that simple substitution chemistry in the anthraquinone series is more complicated than might previously have been thought.

5 Self-Assembling Systems

The definition of "self-assembly" is difficult to write because of the many possible views of the phenomenon. The reaction of a proton with an anion is self-assembly but it is usually referred to as the chemical process of protonation. The assembly of two carboxylic acids to form a dimer is certainly assembly. Likewise, the association of adenine and thymine to form the base pair present in deoxyribonucleic acid is assembly. All of these interactions are readily comprehended in simple structural terms, even though the phenomena themselves are complex. The situation is different for amphiphiles. Compounds having a polar end (head) and an oleophilic end (tail) tend to assemble but the forms are numerous and diverse. In the present article, we will deal with several forms of assembly. We have developed systems that can dimerize in a predictable fashion and we have prepared molecules that serve as monomers for the formation of micelles, niosomes, or vesicles.

5.1 Steroidal Lariat Ethers

Much thought has been devoted by many workers to mechanisms by which macrocyclic polyethers may be ordered into channels. This is, in a way, an obvious application for these molecules since they are cylindrical, they have available positions to which bonds may be connected, and the resulting cylinders may be cation complexing. At the outset of the effort described below, our thought was to use the known ordering properties of the steriod nucleus to bring molecular organization to a group of molecules that also contained a macrocyclic polyether. In principle, if the steroids stacked as they are known to do in liquid crystalline phases [39], that same organization would be enforced upon the connected macrocycle. We further reasoned that if there was a change in the organization of the steroidal aggregate due to an alteration in temperature, the degree of macroring overlap in the various layers might be affected. As the overlap changed, the permeability of the channel would also change. In effect, this membrane's permeability would be tunable whereas a system in which only "open" or "closed" was possible is a binary switch.

Unfortunately, too little information is yet available from our laboratory to decide if this concept is feasible. The steroidal lariat ethers are amphiphilic, however, and able to undergo association in more traditional ways. The macrocyclic polyether ring is polar and even more polar when complexing a cation. The cholesteryl sidechain contains 27 carbon atoms and is obviously lipophilic. Indeed, cholesterol is a common component of many natural membrane systems and its presence helps to increase order (rigidity). We thus undertook the syntheses of steroidal lariat ethers so that their many potential properties could be assessed.

5.1.1 Syntheses of Steroidal Lariat Ethers

Our work in the lariat ether area has focused on the carbon-pivot and the nitrogen-pivot molecules. Access to these systems is described above. Several choices needed to be made concerning which group of compounds to use in developing the steroidal lariat ethers. For the carbon-pivot systems, there were questions of ring size, functional group used for attachment, length of connector, and others. For the practical reasons outlined in Sect. 2.3, we favored formation of the 15-membered rings in the carbon-pivot series. The obvious functional groups that could be used for connecting the steroids to the macrorings are ether and ester. An obvious approach to such ethers from 2-tosyloxymethyl-15-crown-5 and cholestanol using a NaH/THF system, failed. Instead of undergoing substitution, the tosylate group eliminated to form 2-vinylidene-15-crown-5 [23].

At the time our early work on these compounds began, 2-carboxy-15-crown-5 was relatively inaccessible to us. Although we were able to develop some oxidative approaches to this molecule, none afforded a high enough yield to make it a reasonable precursor. Thus, we prepared an ether derivative as the first example of this family. Commercially available cholestanol (dihydrocholesterol) was treated with allyl chloride, sodium hydroxide, and tetrabutylammonium bisulfate. The allyl ether was obtained in 72% yield. Oxidation to the glycol (80% yield) was accomplished using a catalytic amount of OsO_4 and *N*-methylmorpholine-*N*-oxide. The diol was then cyclized using NaH and $TsO(CH_2CH_2O)_4Ts$ to afford the ether product in 10% yield [23].

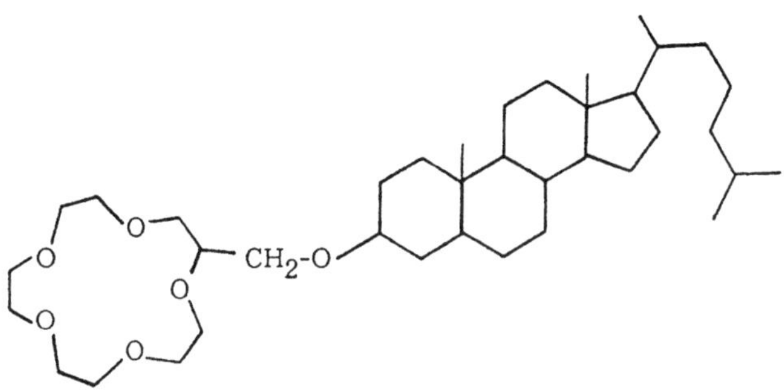

Because the synthesis of this compound was cumbersome and because the yield was modest at best, we turned our attention to the preparation of nitrogen-pivot lariat ether derivatives of steroids. The question of ester *vs* ether linkage was resolved quickly in favor of esters but the carbonyl group could be attached to cholesterol (to form an ester) or to the nitrogen atom to form an amide. An early attempt to prepare a representative of this family used commercially available cholesteryl chloroformate. Thus, cholesteryl chloroformate, aza-15-crown-5, and triethylamine were heated at 90 °C in DMF for 48 hours. After chromatography and recrystallization, the carbamate crown˙was obtained in 34% yield [23].

Although it appeared from our examination of CPK molecular models that this compound might be too rigid to give interesting results in aggregation experiments, it proved a suitable candidate for X-ray crystallographic analysis. The structure was obtained at Louisiana State University in collaboration with Profs. Frank Fronczek and Richard Gandour [23]. It proved interesting in the sense that no structure of a 15-membered crown ether containing only carbon, oxygen, and nitrogen had previously been reported. The key feature of this structure is that the a ring methylene is turned inward to partially fill the macroring hole. The solid state structure was shown in Fig. 1, above.

Most of the work accomplished to date on these interesting macrocyclic systems involves formation of an activated derivative of the steroid which is then coupled to the macrocyclic amine. Typically, cholesterol and triethylamine in benzene was treated with chloroacetyl chloride to afford cholesteryl chloroacetate. Alkylation using this derivative proved facile because the primary chloride is activated by the adjacent carbonyl group. Yields for the alkylation of an aza-crown by cholesteryl chloroacetate were typically 60% or higher [23] making these molecules readily accessible.

5.1.2 Aggregation of Steroidal Lariat Ethers

When a typical steroidal lariat ether, such as the one shown directly above, was melted and dispersed in deionized water (concentration 1–5 mM), and then sonicated, the turbid dispersion gradually clarified until a plateau was reached. The sonicated solutions were filtered through Nuclepore polycarbonate membranes of 0.2 millimicron porosity. The pH of the system remained about 8.5 throughout this process The resulting solution was stable for weeks and had the opalescent and bluish color characteristic of vesicles [40]. These vesicles were

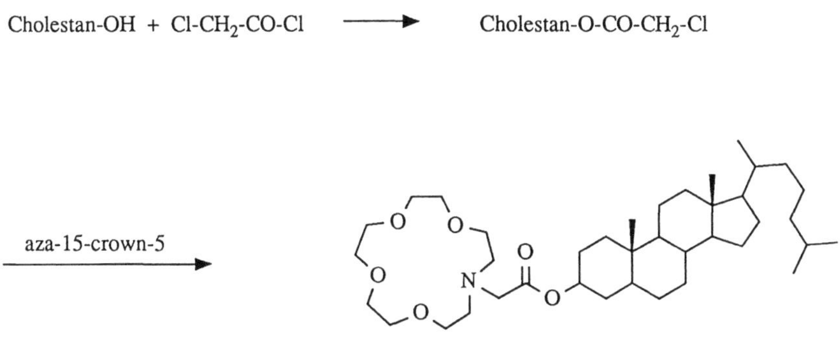

Scheme 2. Synthesis of *N*-pivot, steroidal lariat ether

characterized in a variety of ways. Transmission electron microscopy using metallic staining agents showed the pattern typical of unilamellar vesicles. Size was also assessed using the Coulter Model N4-SD instrument and the "size distribution processor" program. Light scattering data were obtained for the two aza-15-crown-5 derivatives shown below. The average niosome diameter was found to be 30–33 nm for both systems, the unsaturated derivative exhibiting both a slightly larger average size and slightly larger particle range than the unsaturated one.

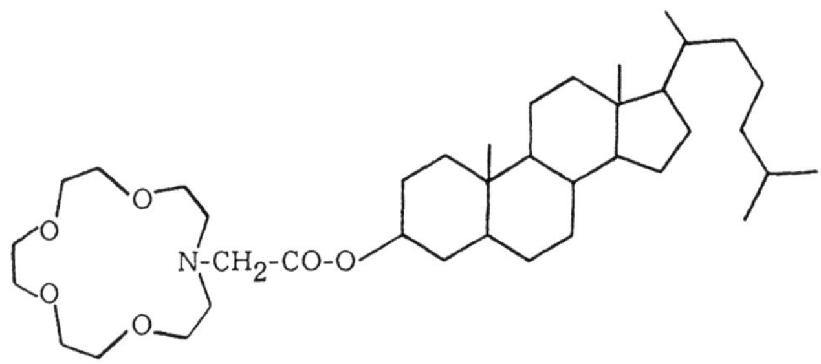

Addition of 0.1 *M* lithium, sodium, or potassium chloride to the solution did not alter the form of the aggregates except that the niosomes were transformed, formally at least, into vesicles.

The behavior of the 18-membered ring, cholesteryl lariat ether proved somewhat different from the 15-membered ring compounds described above [41]. *N*-(3-Cholesteryloxycarbonylmethyl)aza-18-crown-6 formed micelles rather than niosomes in water. When no cation was present in the solution or when weakly-bound Li^+ was present, unusual cloud point behavior was observed. A cloud point near 64 °C was observed in either case and it was relatively sharp. When either Na^+ or K^+ was present in the solutions, cloudiness was observed on heating to 74 °C or 89 °C, respectively, but the solution remained cloudy on cooling until it reached 22 °C and 34 °C respectively. This unusual behavior was reproducible many times using the same sample solutions. This hysteresis may well be due to the formation of a lyotropic mesophase that reaches equilibrium slowly.

5.1.3 Rigidity of Steroidal Lariat Ether Membranes

The results described above are both intriguing and gratifying. The steroidal lariat ethers aggregate in the complete absence of cations to give stable niosomes. So far as we are aware, there is only one other example extant of niosome formation from synthetic monomers in the absence of added cholesterol [42]. The addition of cholesterol, patterned after its appearance in many natural membrane systems, causes enhanced rigidity. In our case, no cholesterol was added but the steroid is obviously an integral part of the membrane assembly. We sought to assess the rigidity of these systems using the electron paramagnetic resonance (EPR) technique [43].

Nitroxide spin labels have been used extensively in biological systems to assess molecular motions and interactions [44]. The radicals may be covalently linked to the system or they may be intercalated into a membrane. In either case, the coupling constants and correlation times give a measure of system mobility (rigidity) [45].

The three steroidal lariat ether systems (15u for unsaturated, 15s for saturated, and 18u) illustrated and discussed above were examined using the steroidal nitroxide probe shown below. Niosomes of 15u, 15s, and 18u in the presence of nitroxide exhibited coupling constants of 32.1 G, 32.6 G, and 32.0 G. The coupling constant 33.4 G is the limiting value observed for the frozen dispersion of any of these. This remarkable rigidity is confirmed by correlation times [45] and order parameters (S [46]). We estimate from these data that our synthetic niosomes are about 300-fold *less* mobile than egg lecithin vesicles.

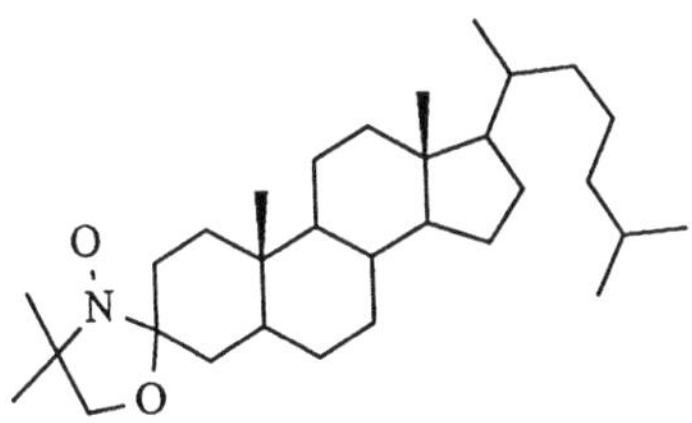

5.2 Liopophilic Cryptands

The essential property of the lipophilic lariat ether compounds discussed here is that they have a polar head group and a lipophilic tail. These compounds are unique in having steroidal tails but other surface-active crown ether compounds have been prepared and studied by Kuwamura [47], Okahara [48], Turro [49], and in other groups as well [50]. Other, potentially surface active, compounds have also been prepared although their surface activity had not, to our knowledge, been assessed prior to our work in this area [23,40,41].

Montanari, Tundo, and their coworkers [51] produced several interesting, lipophilic complexing agents as part of their extensive studies in the field of phase transfer catalysis [52]. Phase transfer catalysis relies on the ability of the catalyst to ion-pair with a nucleophile and conduct it from water into an organic phase where reaction occurs. Montanari demonstrated that phase transfer catalytic activity could be enhanced by complexing the cationic portion of the reactant salt using a lipophilic crown or cryptand. The same compounds used by Montanari and coworkers suggested themselves for study as possible surface active systems.

The lipophilic cryptands of Montanari in general and the previously unknown steroidal cryptands [53] in particular were of interest to us. When the monomer's lipophilic sidearm is a hydrocarbon sidechain, the CPK molecular models do not show the aesthetically pleasing stacking arrangement observed for the steroidal systems. The cryptands exhibit very strong and sometimes quite selective cation binding properties [54]. We already knew from the work of Evans [55] and Kaifer [56] that even when no sidechain was present, the cryptands could modify micelles formed from sodium dodecyl sulfate.

2-*n*-Tetradecyl-[2.2.2]-cryptand ([2.2.2.]-C_{14}) was prepared by the route previously described in the literature [51]. The analogous steroidal compound ([2.2.2]-cholesteryl) was prepared by a variation of the Montanari approach in which cholestanyl allyl ether was converted to the diol by oxidation. This diol was, in turn, converted into the dicarboxylic acid derivative required for reaction with diaza-18-crown-6.

Cholestanyl-OH-(allyl chloride) ⟶ Cholestanyl-O–$CH_2CH{=}CH_2$

(oxidative hydroxylation) ⟶ Cholestanyl-O–CH_2–CH(OH)–CH_2OH

+ ($ClCH_2COOH$/base)

⟶ Cholestanyl-O–CH_2–CH(CH_2COOH)–CH_2–OCH_2COOH

The diacid was then converted into the diacyl chloride ($SOCl_2$) and then cyclized with 4,13-diaza-18-crown-6 to give the steroidal cryptand as a diamide. Reduction of the amide functions was accomplished using diborane. The cryptand product is illustrated below.

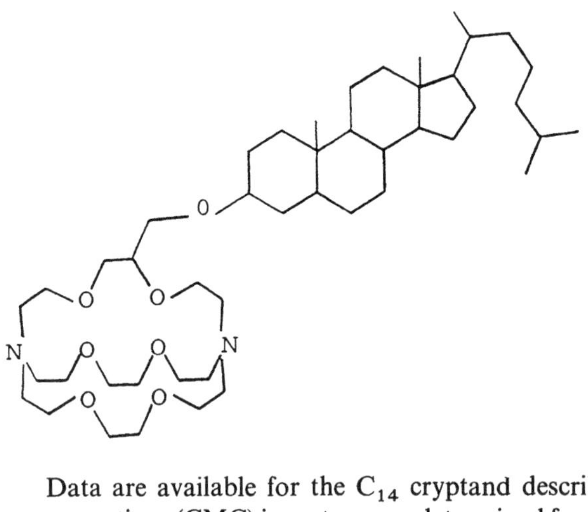

Data are available for the C_{14} cryptand described above. Critical micelle concentrations (CMC) in water were determined for these systems using a sudan III dye solubilization technique or using a surface tensiometer [41]. The critical micelle concentrations for the C_{14}-cryptand ([2.2.2]-C_{14}) were all in the range 0.10–0.14 mM in the absence of any added salt or in the presence of one equivalent of LiCl, NaCl, or KCl. When $AgNO_3$ was present, the CMC about doubled (0.24) but when $BaCl_2$ was added, the CMC increased to 1.31. Likewise, the cloud points (CPs) observed for these compounds ranged from 12.5–41 °C in the presence of an alkali metal chloride salt. When either $AgNO_3$ or $BaCl_2$ was added, the cloud point increased to more than 82 °C.

An interesting observation resulted from a study in which KCl was added incrementally to [2.2.2]-C_{14}. We expected that the presence of K^+ ion would influence both the CMC and CP for these systems. We anticipated that whatever change took place, the alteration would correspond to salt concentration until 1.0 equivalents of K^+ was added. We were surprised to discover that as the ratio of cryptand/KCl was varied from 1:0.2 to 1:5, the cloud point varied from 32.0 °C to 89.0 °C. When one equivalent each of cryptand and potassium cation were present, the CP was 41.0 °C.

The obvious expectation is that whatever influence a cation such as K^+ would have on this system, would be maximized at 1.0 equivalent of cation. This is because the equilibrium constant for the reaction

$$K^+_{(water)} + [2.2.2]_{(water)} \longrightarrow \text{complex}_{(water)}$$

is more than 250,000 in favor of complexation [59]. We feel that complexation must surely occur, but that as a larger and larger number of cryptand head groups complex cations, the surface charge increases to the point that incremental complexation becomes very costly in energy terms [57].

Most of the steroidal lariat ethers studied thus far form niosomes or vesicles when salts are added. *N*-(3-Cholesteryloxycarbonylmethyl)-aza-18-crown-6, in contrast, forms micelles. The cryptands studied thus far also form micelles both in the presence and absence of cations. The difference between the cryptand and lariat ether systems can be accounted for readily by saying that the cryptand head groups are larger, have more donors, are more polar, are probably more hydrated, and are stronger cation binders. These arguments cannot all be made for the 18-membered cholesteryl lariat ether we have studied, but it forms micelles rather than niosomes. To be sure, the head to tail size ratio suggested by Evans [58] differs for the 15- and 18-membered ring lariat ethers as it does for the cryptands, but such as correlation, no matter how useful, is largely empirical.

A *bis*(steroidal) derivative of diaza-18-crown-6 has recently been brought to hand and its aggregation properties (if any) will be studied. Its structure is shown below.

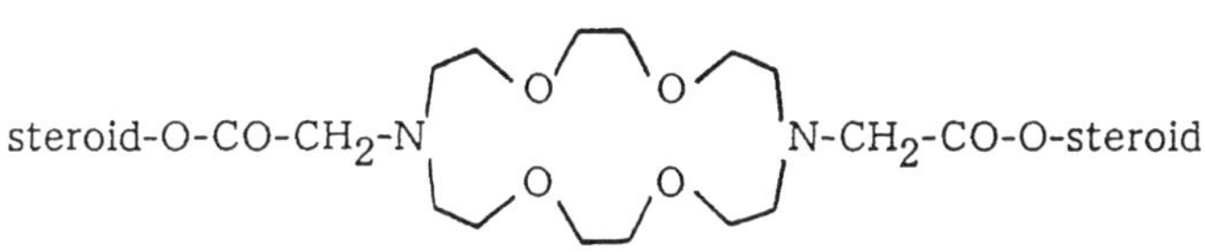

Many questions remain to be answered about this structure. For example, will it be too crowded for the two steroid residues to align? Will the connecting chains be long enough to permit alignment while allowing the head group to be solvated? What will be the effect of cations? Of course, a key question for all of these systems is whether micelles or niosomes will form. Experiments addressing these and other questions are underway and likely to continue for some time.

5.3 Hydrogen-Bonded Systems

One goal of the work underway is to use weak or "feeble" molecular forces to augment covalent bonds in the formation of novel structures and complexes. Nature, for example, forms the required backbone of a peptide using the so-called primary structure. The order of amino acids is crucial to both the form and function of the peptide or enzyme. The particular order is a necessary, but not sufficient, condition. The active conformations of an enzyme or peptide are ensured by much weaker interactions such as conformational interactions, salt bridges, hydrogen bonds, etc.

An excellent example of how important these "feeble" forces can be is found in the molecule valinomycin [4]. As noted above, nature has used primary structure to alternate amino and hydroxy acids in this molecule. This has the consequence that more polar amide and less polar ester carbonyl groups alternate. Nature has used "feeble" forces in the conformation of this molecule: by alternating D- and L-amino acids, a folding of valinomycin is accomplished without the rigidity that accompanies insertion of proline. Finally, the folded or pre-binding [4] conformation of valinomycin is stabilized by six hydrogen bonds involving the amide carbonyl groups. This not only stabilizes the folded structure, it occupies the more polar amide carbonyl groups leaving an octahedral array of ester carbonyl groups in position to bind potassium cation. The latter ion is less charge dense than Na^+, Mg^{2+}, or Ca^{2+}, the other ions common in serum [60], and this enhances selectivity for K^+.

In order to see if appropriately placed hydrogen bonding forces could be used to stabilize complex structures, we designed the following system based upon the diaza-18-crown-6 framework we [20, 21] and others [22] have extensively explored. Our original notion was to prepare a 4,13-diaza-18-crown-6 derivative having a purine on one side and a pyrimidine on the other, both attached to nitrogen atoms through a $(CH_2)_n$ chain. In principle, a compound having adenine on one sidearm and thymine on the other, represented schematically as A-O-T, could dimerize using the same hydrogen bonding forces known to stabilize the dimer structure of DNA. In this shorthand, A stands for adenine, T stands for thymine, - represents a connecting chain, and O represents a macrocycle. Since the monosubstitution of a diaza-18-crown-6 proved a daunting prospect, we turned our attention instead to the synthesis of two precursors, A-O-A and T-O-T.

The preparation of these materials has been outlined in a recent communication [61]. Osmometric molecular weight determinations showed that the dimer structure shown below was favored in water in which solvent these measurements were taken. Evidence was also obtained for a further association with a *bis*-(ammonium salt) although, for reasons discussed elsewhere [59], the extent of aggregation remains difficult to assess. The structure of the presumed complex is shown below. At this writing, an effort is underway to further explore this complexation chemistry and to extend it to the cytosine/guanine analogs as well as to the more challenging A-O-T structure.

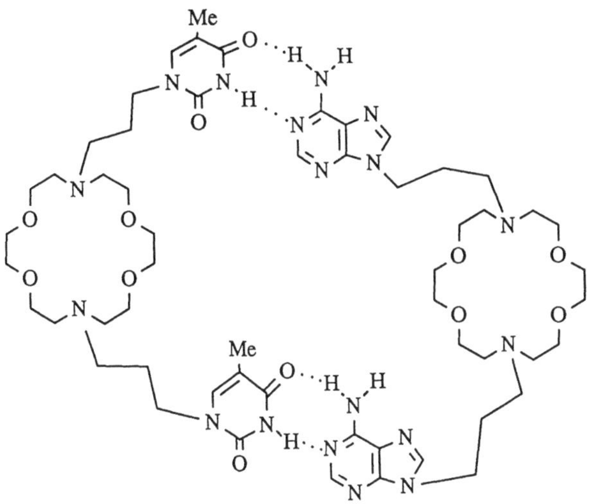

6 Overview and Conclusions

The lariat ethers are compounds based on a concept of three-dimensionality coupled with flexibility. Cations require three-dimensional solvation and this is often provided by a combination of ligand and solvent. In the lariat ethers, a complete solvation sphere may be provided by a combination of ring and sidearm. Although the single-armed compounds sometimes include solvent in the coordination sphere, the sidearms serve this function in the bibracchial systems for every case of which we are aware. Both cation binding strength and selectivity may be controlled in lariat ethers although it is a complex process.

The lariat ethers may be used in ways that go far beyond simple cation complexation. We have already demonstrated that electrochemically reduced systems can be used to enhance cation transport in model membranes. We have further demonstrated that membranes themselves may be constructed from lariat ethers and these systems have interesting and unusual properties. Notable among the latter is the remarkable rigidity of steroidal lariat ether membranes. At the other end of the size spectrum, we have shown that organized assemblies may be obtained from as few as two, appropriately-substituted lariat ether monomers.

It now remains for us to combine these individual properties. We look forward to cation transport agents that can be switched, pumped, and otherwise subtly controlled to alter selectivity, transport rate, and lipophilicity. More complex and larger assemblies may be possible and these may include a synthetic membrane that is itself switchable. Using these principles, we may be able to control membrane permeability, selectivity, and other properties.

Acknowledgements. We warmly thank the many coworkers and collaborators with whom it has been our privilege to work. These include the undergraduate, graduate, and post-doctoral associates whose names may be found on cited literature. We are especially indebted to Professor Angel Kaifer, former student and now fine colleague for his early work in many portions of these studies. Jerry L. Atwood, the late Jim Christensen, Edward Eyring, Frank R. Fronczek, Richard D. Gandour, Reed Izatt, Angel Kaifer, Charles R. Morgan, Mitsuo Okahara, Sergio Petrucci, and the late Iwao Tabushi have all proved to be fine friends and excellent collaborators. Professor Donald J. Cram deserves special thanks for his inspiration. Finally, we thank W. R. Grace and Co., Bendix corporation, Lion Detergent Corp., and especially the National Institutes of Health (grants to GWG and LE) for supporting portions of the work described here.

7 References

1. Westley JW (1982) Polyether antibiotics, Marcel Dekker, New York
2. Ovchinnikov YA, Ivanov VT, Shkrob AM (1974) Membrane-active complexones, Elsevier, Amsterdam
3. Dobler M (1981) Ionophores and their structures, Wiley Interscience, New York, p 48.
4. a. Grell E, Funck T, Eggers F (1975) F. In: Eisenman G, (ed.) Membranes, Marcel Dekker, New York, vol. 3, p 1
   b. Liesegang GW, Eyring EM (1978) In: Izatt RM, Chrisensen JJ (eds) Synthetic multidenate Macrocycle Compounds, Academic, New York, 289
5. a. Lehn J-M, Sauvage JP, (1975) J. Am. Chem. Soc. 97: 6700
   b. Cox BG, Schneider H, Stroka J (1978) J. Am. Chem. Soc. 100: 4746
   c. Yee EL, Tabib J, Weaver MJ (1979) J. Electroanal. Chem. 96: 241
6. Truter MR (1973) Structure and Bonding 16: 71
7. Stryer L (1981) Biochemistry, 2nd edn, Freeman, p 774
8. a. Liesegang GW, Farrow MM, Rodriguez LJ, Burnham RK, Eyring EM, Purdie N (1978) Int. J. Chem. Kin. 10: 471
   b. Live D, Chan SI (1976) J. Am. Chem. Soc. 98: 3769
9. Gresser P, Boyd DW, Albrecht-Gary AM, Schwing JP (1980) J. Am. Chem. Soc. 102: 651
10. Gisin BF, Merrifield RB, Tosteson DC (1969) J. Am. Chem. Soc. 91: 2691
11. Shenyakin MM, Ovchinnikov YA, Ivanov VT, Antonov VK, Vinogradova EI, Shkrob AM, Estratov AV, Malenkov GG, Laine IA, Melnik EI, Ryabova ID (1969) J. Membrane Biol. 1: 402
12. a. Gokel GW, Dishong DM, Diamond CJ (1980) J. Chem. Soc. Chem. Commun. 1980: 1053
    b. Gokel GW, Dishong DM, Diamond CJ (1981) J Tetrahedron Lett. 1981: 1663
13. a. Pedersen CJ (1967) J. Am. Chem. Soc. 89: 2495
    b. Pedersen CJ (1967) J. Am. Chem. Soc. 89: 7077
14. a. Michaux G, Reisse J (1982) J. Am. Chem. Soc. 104: 6895
    b. Arnold KA, Echegoyen L, Gokel GW (1987) J. Am. Chem. Soc. 109: 3713
15. Gold V, Sghibartz CM (1978) J. Chem. Soc. Chem. Commun. 1978: 507
16. Dishong DM, Diamond CJ, Cinoman MI, Gokel GW (1983) J. Am. Chem. Soc. 105: 586
17. a. Nakatsuji Y, Nakamura T, Okahara M, Dishong DM, Gokel GW (1982) Tetrahedron Lett. 23: 1351
    b. Nakatsuji Y, Nakamura T, Okahara M, Dishong DM, Gokel GW (1983) J. Org. Chem., 48: 1237
    c. Nakatsuji Y, Nakamura T, Yonetani M, Yuya H, Okahara M (1988) J. Am. Chem. Soc. 110: 531
18. a. Schultz RA, Dishong DM, Gokel GW (1982) J. Am. Chem. Soc. 104: 625
    b. Schultz RA, White BD, Dishong DM, Arnold KA, Gokel GW (1985) J. Am. Chem. Soc. 107: 6659
19. Kuo PL, Miki M, Ikeda I, Okahara M (1978) Tetrahedron Letters 1978: 4273
20. Gatto VJ, Gokel GW (1984) J. Am. Chem. Soc. 106: 8240

21. a. Gatto VJ, Arnold KA, Viscariello AM, Miller SR, Gokel GW (1986) Tetrahedron Lett. 27: 327
b. Gatto VJ, Arnold KA, Viscariello AM, Miller SR, Morgan CR, Gokel GW (1986) J. Org. Chem. 51: 5373
22. a. Ricard A, Capillon J, Quivoeron C (1985) Polymer 25: 1136
b. Tsukube H (1984) Bull. Chem. Soc. Jpn. 57: 2685
c. Tsukube H (1984) J.C.S. Chem. Commun. 1984: 315
d. Tsukube H (1983) J.C.S. Chem. Commun 1983: 970
e. Keana JFW, Cuomo J, Lex L, Seyedrezai SE (1983) J. Org. Chem. 48: 2647
f. DeJong FA, Van Zon A, Reinhoudt DN, Torny GJ, Tomassen HPM (1983) Recl. J. R. Neth. Chem. Soc. 102: 164
g. Shinkai S, Kinda H, Araragi Y, Manabe O (1983) Bull. Chem. Soc. Jpn. 56: 559
h. Kobayashi H, Okahara M (1983) J.C.S. Chem. Commun. 1983: 800
i. Bogatsky AV, Lukyanenko NG, Pastushok VN, Kostyanovsky RG (1983) Synthesis 1983: 992
j. Frere Y, Gramain P (1982) Makromol. Chem. 183: 2163
k. Tazaki M, Nita K, Takagi M, Ueno K (1982) Chemistry Lett. 1982: 571
l. Cho I, Chang S-K (1980) Bull. Korean Chem. Soc. 1980: 145
m. Gramain P, Kleiber M, Frere Y (1980) Polymer 21: 915
n. Kulstad S, Malmsten LA (1979) Acta Chem. Scand., Ser. B B33: 469
o. Wester N, Voegtle F (1978) J. Chem. Res. (S) 1978: 400
p. Takagi M, Tazaki M, Ueno K (1978) Chemistry Lett. 1978: 1179
23. Gokel GW, Hernandez JC, Viscariello AM, Arnold KA, Campana CF, Echegoyen L, Fronczek FR, Gandour RD, Morgan CR, Trafton JE, Miller SR, Minganti C, Eiband D, Schultz RA, Tamminen M (1987) J. Org. Chem. 52: 2963
24. Gokel GW, Goli DM, Minganti C, Echegoyen L (1983) J. Am. Chem. Soc. 105: 6786
25. Miller SR, Cleary TP, Trafton JE, Smeraglia C, Fronczek FR, Gandour RD, Gokel GW (1989) J. Chem. Soc. Chem. Commun., 808
26. a. Davidson RB, Izatt RM, Christensen JJ, Schultz RA, Dishong DM, Gokel GW (1984) J. Org. Chem. 49: 5080
b. Arnold KA, Echegoyen L, Fronczek FR, Gandour RD, Gatto VJ, White BD, Gokel GW (1987) J. Am. Chem. Soc. 109: 3716
c. Arnold KA, Trafton JE, Gokel GW (1990) unpublished
27. Arnold KA, Viscariello AM, Kim M, Gandour RD, Fronczek FR, Gokel GW (1988) Tetrahedron Lett. 1988: 3025
28. a. Calverley MJ, Dale J (1982) Acta Chem. Scand., B. 36: 241
b. Krane J, Amble E, Dale J, Daasvatn K (1980) Acta Chem. Scand., B. 34: 255
c. White BD, Dishong DM, Minganti CM, Arnold KA, Goli DM, Gokel GW (1985) Tetrahedron Lett. 26: 151
d. White BD, Arnold KA, Garrell RL, Fronczek FR, Gandour RD, Gokel GW (1987) J. Org. Chem. 52: 1128
e. Arnold KA, Mallen J, Trafton JE, White BD, Fronczek FR, Gehrig LM, Gandour RD, Gokel GW (1988) J. Org. Chem. 53: 5652
29. a. Echegoyen L, Gokel GW, Kim MS, Eyring EM, Petrucci S (1987) J. Phys. Chem. 1987: 3854
b Gokel GW (1987) Biophysical Chemistry 26: 225
30. a. Stolwijk TB, Sudholter EJR, Reinhoudt DN (1987) J. Am. Chem. Soc. 109: 7042
b. Goddard JD (1985) J. Phys. Chem. 89: 1825
c. Stein WD (1986) Transport and diffusion across cell membranes, Academic, New York
31. Geske DH, Ragle JL, Bambenek MA, Balch AL (1964) J. Am. Chem. Soc. 86: 987
32. a. Kaifer A, Echegoyen L, Gustowski D, Goli DM, Gokel GW (1983) J. Am. Chem. Soc. 105: 7186
b. Gustowski DA, Echegoyen L, Goli DM, Kaifer A, Schultz RA, Gokel GW (1984) J. Am. Chem. Soc. 106: 1633
c. Morgan CR, Gustowski DA, Cleary TP, Echegoyen L, Gokel GW (1984) J. Org. Chem. 49: 5008
d. Kaifer A, Gustowski DA, Echegoyen L, Gatto VJ, Schultz RA, Cleary TP, Morgan CR, Rios AM, Gokel GW (1985) J. Am. Chem. Soc. 107: 1958
e. Gokel GW, Echegoyen L: United States Patent Number 4,631,119, December 23, 1986
33. a. Kulstad S, Malmsten LA (1980) J. Inorg. Nucl. Chem. 42: 573
b. Kolthoff IM, Chantooni MK (1980) Anal. Chem. 52: 1039

c Gutknecht J, Schneider H, Stroka J (1978) Inorg. Chem. 17: 3326
d. Kolthoff IM, Chantooni MK (1980) Proc. Natl. Acad. Sci. U.S.A. 77: 5040
34. a. Echegoyen L, Gustowski DA, Gatto VJ, Gokel GW (1986) J. Chem.'Soc. Chem. Commun. 1986: 220
b. Gustowski DA, Delgado M, Gatto VJ, Echegoyen L, Gokel GW (1986) Tetrahedron Lett. 1986: 3487
c. Gustowski DA, Delgado M, Gatto VJ, Echgeoyen L, Gokel GW (1986) J. Am. Chem. Soc. 108: 7553
d. Echeverria L, Delgado M, Gatto VJ, Gokel GW, Echgeoyen L (1986) J. Am. Chem. Soc. 108: 6825
e. Delgado M, Gustowski DA, Yoo HK, Gatto VJ, Gokel GW (1988) J. Amer. Chem. Soc. 110: 119
f. Echegoyen L, Gokel GW, Echegoyen LE, Chen Z-H, Yoo HK (1988) J. Inclusion Phenom. 7: 257
35. a. Voegtle F, Weber E (1974) Angew. Chem. Int. Ed. Engl. 13: 814
b. Voegtle F, Weber E (1979) Angew. Chem. Int. Ed. Engl. 18: 753
36. a. Shannon RD (1976) Acta Crystallogr., A 32: 751
b. Henderson P (1982) Inorganic geochemistry, Pergamon, New York
37. Echeverria L, Delgado M, Gatto VJ, Gokel GW, Echegoyen L (1986) J. Am. Chem. Soc. 108: 6825
38. a. Yoo HK, Davis DM, Chen Z, Echegoyen L, Gokel GW, Tetrahedron Letters (1990) in press.
b. Lu T, Yoo HK, Zhang H, Bott S, Atwood JL, Echegoyen L, Gokel GW, J. Org. Chem. (1990), in press
39. a. Sisido M, Takeuchi K, Imanishi Y (1984) J. Phys. Chem. 88: 2893
b. Hornreich RM, Shtrikman S (1984) Phys. Rev. A 29: 3444
40. Echegoyen LE, Hernandez JC, Kaifer A, Gokel GW, Echegoyen L (1988) J. Chem. Soc. Chem. Commun. 1988: 836
41. Echegoyen LE, Portugal L, Miller SR, Hernandez JC, Echegoyen L, Gokel GW (1988) Tetrahedron Letters 4056
42. a. Baillie AJ, Florence AT, Hume LR, Muirhead GT, Rogerson A (1985) J. Pharm. Pharmacol. 37: 863
b. Ribier A, Handjani-Vila RM, Bardez E, Valeur B (1984) Colloids and Surfaces. 10: 155
c. Azmin MN, Florence MN, Handjani-Vila RM, Stuart JFB, Vanlerberghe G, Whittaker JS (1985), J. Pharm. Pharmacol 37: 237
43. Fasoli H, Echegoyen LE, Hernandez JC, Gokel GW, Echegoyen L (1989) J. Chem. Soc. Chem. Commun. 578
44. a. Stone TJ, Buchman T, Mordio RL, McConnell HM (1965) Proc. Nat. Acad. Sci., USA 54: 1010
b. Hamilton CL, McConnell HM (1968) Spin labels. In: Rich A, Davidson N (eds) Structural chemistry and molecular biology, W.H. Freeman, San Francisco
c. McConnell HM, McFarland BJ (1970) Quart. Rev. Biophys., 3: 91
d. Seelig T (1970) J. Am. Chem. Soc. 92: 3881
e. Presti FT, Chan SI (1982) Biochemistry 21: 3821
f. Hubell W, McConnell HM (1969) Proc. Nat. Acad. Sci. USA 64: 20
g. Hubell W, McConnell HM (1971) J. Am. Chem. Soc. 93: 314
h. Ojil Marrot JB, Roux M, Mamin L, Favre E, Jhaux PH (1987) Biochem. Biophys. Acta 897: 341
i. Campbell JD, Dwek RA (1984) Biological spectroscopy, Benjamin/Cummings, Menlo Park
j. Knowles PF, Marsh D, Rattle HWE (1976) Magnetic resonance of biomolecules, John Wiley, New York
k. Nordio PL (1976) In: Berliner LJ (ed) Spin labeling, Academic, New York, p 5
45. Freed J (1976) In: Berliner LJ (ed) Spin labeling, Academic, New York, p 53
46. a. McFarland BG, McConnell HM (1971) Proc. Nat. Acad. Sci. (U.S.A.), 68: 1274
b. Haering G, Luisi PL, Hauser H (1988) J. Phys. Chem. 92: 3574
47. a. Kuwamura T, Kawachi T (1979) Yukagaku 28: 195 (Chem. Abstr. 90: 206248d)
b. Kuwamura T, Akimaru M, Takahashi HL, Arai M (1979) Kenkyu Hokoku-Asahi Garasu Kogyo Gijutsu Shoreikai 35: 45 (Chem. Abstr. 95: 61394q)
c. Kuwamura T, Yoshida S (1980) Nippon Kagaku Kaishi 1980: 427 (Chem. Abstr. 93: 28168e)
48. a. Okahara M, Kuo PL, Yamamura S, Ideda I (1980) J. Chem. Soc. Chem. Commun., 1980: 586
b. Kuo P, Tsuchiya K, Ikeda I, Okahara M (1983) J. Colloid Interface Sci. 92: 463

49. Turro NJ, Kuo P (1986) J. Phys. Chem. 90: 837
50. a. LeMoigne J, Gramain P, Simon J (1977) J. Colloid Interface Sci. 60: 565
b. Morio Y, Pramamuro E, Gratzel M, Pelizzetti F, Tundo P (1979) J. Colloid Interface Sci. 69: 341
51. a. Cinquini M, Montanari F, Tundo P (1977) Gazz. Chim. Ital. 107: 11
b. Landini D, Maia A, Montanari F, Tundo P (1979) J. Am. Chem. Soc. 101: 2526
52. Weber WP, Gokel GW (1977) Phase transfer catalysis in organic synthesis, Springer, Berlin Heidelberg New York
53. Gokel GW, Arnold KA, Delgado M, Echeverria L, Gatto VJ, Gustowski DA, Hernandez J, Kaifer A, Miller SR, Echegoyen L (1988) L. Pure Appl. Chem. 60: 461
54. a. Christensen JJ, Eatough DJ, Izatt RM (1974) Chem. Rev. 74: 351
b. Bradshaw JS (1978) In: Izatt RM, Christensen JJ (eds) Synthetic multidentate macrocyclic compounds, Academic, New York, p 53
c. Lamb JD, Izatt RM, Christensen JJ, Eatough DJ (1978) In: Melsen GA (ed) Coordination chemistry of macrocyclic compounds, Plenum, New York. p 145
d. Izatt RM, Bradshaw JS, Nielsen SA, Lamb JD, Christensen JJ (1985) Chem. Rev. 85: 271
55. Evans DF, Sen R, Warr GG (1986) J. Phys. Chem. 90: 5500
56. Quintela PA, Reno RC, Kaifer AE (1987) J. Phys. Chem. 91: 3582
57. Shamsipur M, Rounaghi G, Popov AI (1980) J. Solution Chem. 9: 701
58. Evans DF, Ninham BW (1986) J. Phys. Chem. 90: 226
59. Gokel GW, Echegoyen L, Kim MS, Hernandez J, DeJesus M (1989) J. Inclusion Phenom. 7: 73
60. Tietz NW, (ed) (1976) Fundamentals of clinical chemistry, 2nd edn, W.B. Saunders, p 879
61. Kim M, Gokel GW (1987) J. Chem. Soc. Chem. Commun. 1987: 1686

On the Way from Small to Very Large Molecular Cavities

Frank Ebmeyer and Fritz Vögtle
Institut für Organische Chemie und Biochemie der Universität Bonn
Gerhard-Domagk-Straße 1, D-5300 Bonn 1, F.R.G.

Design and synthesis of new triply bridged host molecules are described based on the convex/concave complementarity of guest and host. Syntheses have been facilitated through the development and use of a modular synthetic strategy using exchangable building blocks suitable for small as well as for large host cavities. Some of the hosts contain the largest molecular cavities known up till now: They are used successfully for the molecular recognition of functionalized benzene derivatives like 1,3,5-trihydroxybenzene (resorcinol). It is shown that isomeric phenolic guest molecules are selectively discriminated by the new host compounds bearing matching cavities. Complexation of organic guest molecules by enclosure inside the hosts cavities has been possible in water as well as in lipophilic solvents using hosts bearing either solubilizing functions or donor groups.

The modular synthetic concept also allows the construction of strongly complexing macrobicyclic host molecules bearing a small cavity suitable for the highly selective uptake of small metal cations like Fe^{2+}, Fe^{3+}, Rh^{2+}, Ru^{2+}, Ga^{3+}, Cs^{+}. The Fe(III) complexes with a cage type host ("siderophore") containing three catechol units clamped together seem to constitute the most stable complexes encountered in chemistry (complex constant around 10^{59}). Such extremely strong ligands are promising for applications in transition metal complex chemistry, analytical chemistry, in photophysics and photochemistry and in medicinal chemistry.

1 Introduction

The study of molecular recognition processes has been given increasing attention during recent years. Whereas formerly the syntheses of new host structures stood in the foreground, newer investigations concentrate more and more on the selectivity of supramolecular structures, hydrophobic interactions, base pairing in analogy to nucleic acids, hydrogen bridging in the interior of molecular niches, complexation of anions, and highly selective and extremely strong cation complexation. Whilst simple cation complexation by use of crown compounds, podands and cryptands is understood quite well nowadays [1] and is used synthetically [2], more complex synthetic host structures often exhibit surprising properties.

Of special interest in this context are triply bridged macrobicyclic host structures, constructed according to a modular building concept. They allow the construction of tailorshaped host cavities, which encapsulate the guest from all sides almost completely. If intraannular donor functions are introduced into the ligand structure, cavities with varying receptor capabilities are obtained. Such intraannular donor groups consist of O and N containing functions, but unfunctionalized molecular niches allowing hydrophobic interactions are also of interest.

2 Complexation of Metal Cations by Small Host Molecules

2.1 Host Molecules of the Tris-Bipyridine Type

2.1.1 Synthesis

For the synthesis of triply bridged ligands containing bipyridine donor units, mainly carbon-heteroatom bond coupling is used as done throughout the chemistry of macrobicyclic host systems. Thereby alkylation and acylation are applied mostly. These bond connections are often facilitated by an alkali metal

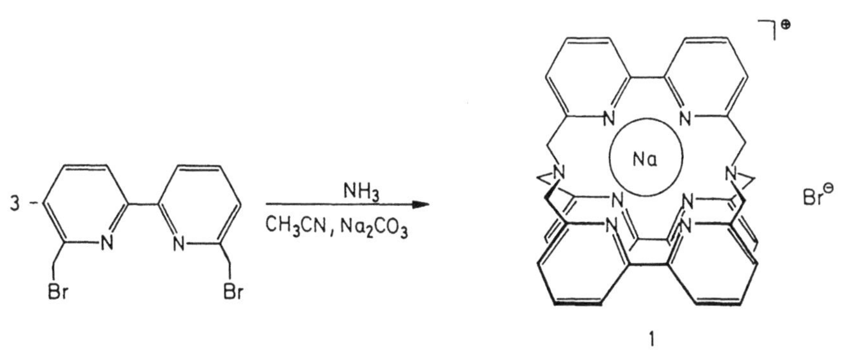

Fig. 1. Template synthesis of the complex **1**

template synthesis. By this way the preparation of the complex **1** starting with 6,6′-bis(bromomethyl)-2,2′-bipyridine succeeds with a yield of 27% [3]: The formation of the sodium complex **1** clearly hints at a template effect. Larger ligands of this type are obtained in the same manner by such template syntheses: Starting with 5,5′-bis(bromomethyl)-2,2′-bipyridine and 1,3,5-tris[(*N*-benzyl)-aminoethyl]benzene the sodium complex of the host **2** is gained with a 4% yield [4]: *N*-acylation is also useful for the preparation of bicyclic *N*-donor-ligands in good yields. The cyclization yet has to start with a corresponding podand. The bicyclic ligand **3** [5] was obtained by a template synthesis using ruthenium as the central ion. During the preparation, the ruthenium stays inside the cavity [6]. It proved impossible to synthesize the ruthenium complex of **3** by introduction of the ruthenium cation into the macrobicyclic ligand **3** directly.

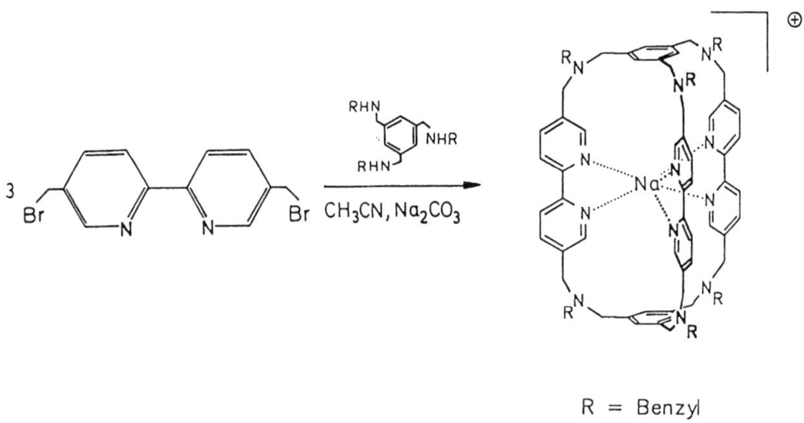

Fig. 2. Template-synthesis of the sodium complex of the bipyridine host **2**

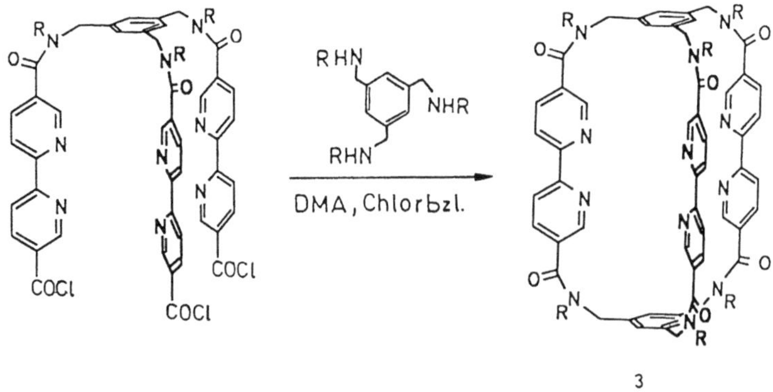

3
R = Benzyl

Fig. 3. Synthesis of the macrobicyclic host **3** with 15% yield

2.1.2 Structure and Properties of Some Alkali Metal- and Transition Metal Complexes

Whereas 2,2′-bipyridine itself forms complexes with alkali metal salts [7] and thereby shows a specific affinity for sodium, with regard to molecular cavities, the size of the niche formed is an additional argument which completes the inherent metal cation selectivity of the *N*-donor-ligand. Apart from the selectivity towards alkali metal cations the stability of complexes also changes. Whereas the simple phenanthroline and bipyridine alkali metal complexes (resp.) in water solution are of low stability, the macrobicyclic ligand systems form complexes of much higher stability. In addition in the 70 eV-electron impact mass spectra of the macrobicyclic complexes of types **1** and **2** intensive signals are observed for the $[M + Na]^+$-ion, which hints to appreciable complex stabilities [3, 4].

Regarding transition metal cations, a further argument for the efficient complexation, namely the orbital topology of the metal cation, has to be taken into account. Thereby, ligands producing distorted ligand-fields for most applications form less useful metal complexes. If an octahedral ligand field is required of the metal, as this is the case with Ru^{2+} and Fe^{2+} then host cavities derived from 5,5′-disubstituted bipyridines often are superior to those derived from 6,6′-disubstituted bipyridines. Dramatic differences are observed regarding the photophysical properties.

The macrobicyclic ligand **3** shows a remarkable stabilisation of Fe^{2+}-ions towards oxidation. For example the Fe^{2+}-complex of **3** is rather stable towards strong oxidants like H_2O_2.

2.1.3 Photophysical Properties of the Ru(II)-Complex

For a long time the intention has existed of using Ru(II)-tris-bipyridine complexes as photosensitizers for relais systems like the well known methyl viologen [8]. Apart from the tendency towards oxidation of the viologens the low photostability of the formerly applied $Ru(bipy)_3^{2+}$ was disturbing. It was shown that the $Ru(bipy)_3^{2+}$-complex after excitation to the first excited singlet-state switches over too quickly to the energetically more favourable ^{3}MLCT-state (MLCT: metal to ligand charge transfer) on account of the strong spin orbit coupling of the electrons of ruthenium. The latter determines the photochemistry of this system. This state can undergo a radiationless deactivation to the ground state on the one hand, and on the other hand dissociate passing a ^{3}MC-state (MC: metal centered) with strongly distorted geometry. It is generally believed that a complex with the coordination number five of the type $Ru(bipy)_2^{2+}$ is formed as an intermediate of this process [9].

There were many attempts to obviate the photodissociation by use of appropriately substituted bipyridines. As a result, the ^{3}MLCT-state was energetically lowered and thereby the radiationless decay was put in the foreground, going along with a dramatic shortening of the lifetime of the ^{3}MLCT-state, thus becoming photochemically unapplicable.

It turns out that a discrimination of the photodissociation and at the same time keeping all other properties of the $Ru(bipy)_3^{2+}$ is only possible by embedding the ruthenium in a macrobicyclic bipyridine ligand, e.g. of type **1**. The Eu^{3+}- and Tb^{3+}-complexes analogous to **1** had proven to be photophysically interesting [10]. Whereas Eu^{3+} and Tb^{3+} do not put high demands on the orbital topology of the ligand field, Ru^{2+} demands a strong octahedral surrounding in order to form a luminescing ^{3}MLCT-state. Otherwise the ^{3}MC-state responsible for the photodissociation sinks energetically under the ^{3}MLCT-state and the photophysical activity is extinguished [11]. The only macrobicyclic Ru^{2+}-bipyridine complex exhibiting the requirements of a tailor-shaped ligand is constituted by the new complex **3**$\cdot Ru^{2+}$. This ruthenium complex exhibits the doubled lifetime of the excited triplet state as does $Ru(bipy)_3^{2+}$ but at the same time has a higher photostability by a factor of 10^4 [12]: The bipyridine ligands substituted in 5,5′-position by carbonyl groups were expected originally to shorten the lifetime of the ^{3}MLCT-state as found in studies of the diethyl-2,2′-bipyridine-5,5′-dicarboxylate [12]. This is not found experimentally due to the orthogonality of the carbonyl groups and the bipyridine planes.

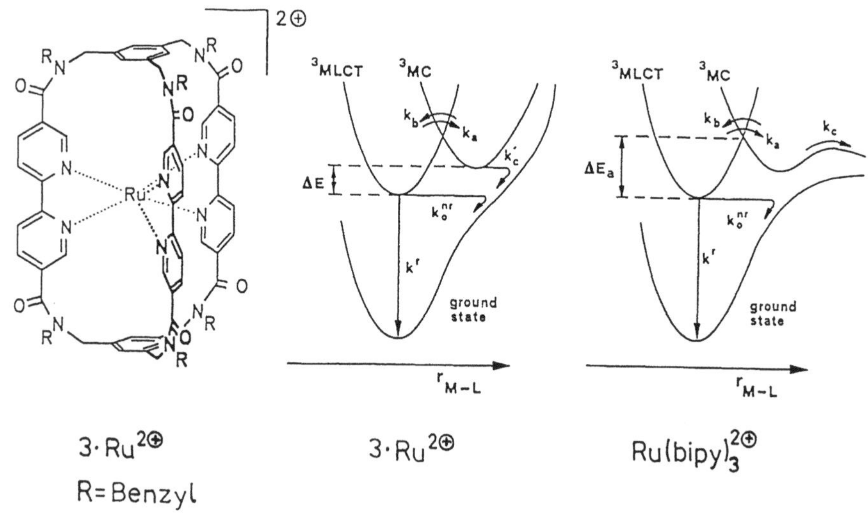

Complex	Absorption	Emission				Photochemistry
	298 K	298 K		90 K		298 K
	λ_{max} [nm] (ε)	λ_{max} [nm]	τ [μs]	λ_{max} [nm]	τ [μs]	Φ_p
$Ru^{II}(bipy)_3$	452 (13000)	615	0.8	582	4.8	$1.7 \cdot 10^{-2}$
$Ru^{II} \cdot 3$	455 (10400)	612	1.7	597	4.8	$< 10^{-6}$

Fig. 4. Photophysical data of the **3** Ru-complex in comparison to the $Ru(bipy)_3^{2+}$-complex

2.2 Host Molecules of the Tris-Catechol Type

2.2.1 Synthesis

The syntheses of triply bridged ligands of the catechol type succeeded with sufficient yields in all cases by threefold amide-bond formation starting with the corresponding functionalised podands, whereas the phenolic hydroxyl groups are protected as OCH_3. One-pot-syntheses with the formation of six amide-bonds have been carried out successfully, but usually give only minor yields. As an example the synthesis of the first representative of such ligands, which was carried out with a yield of 13%, is shown in Fig. 5 [13]: A short time ago a large number of such triply bridged host species were made accessible through a variation of the spacer groups, whereby tris(2-aminoethyl)amine was also used [14]. The amide hydrogens were replaced by lipophilic, exocyclic residues [15]. Approximately at the same time the synthesis was successfully carried out using a Fe^{3+}-template-effect [14a].

2.2.2 Structures and Properties of the Complexes with Metal Cations of Main Group III and Subgroup VIII of the Periodic System

The interest in macrobicyclic ligands of the tris-catechol type resulted from the demand for synthetic analogs of natural siderophores (Greek: iron carriers) such as "enterobactin". This substance isolated from e.g. *Escherichia coli* serves in nature as a Fe^{3+}-carrier and exhibits an extremely high, which was for a long time unequalled, complex constant of $10^{52}\,l\,mol^{-1}$. Experiments were undertaken to obtain synthetic analogs of enterobactin with a similar structure, but with abandonment of the chirality. Was it possible to create new physiologically active siderophore type ligands? It was a question which could be answered positively [16]: MECAM is indeed physiologically active, though it exhibits a lower complex formation constant of $10^{46}\,l\,mol^{-1}$ compared to enterobactin. Like enterobactin itself, MECAM is decomplexed at low pH values. A dramatic improvement of the complexation properties towards Fe^{3+} has been achieved

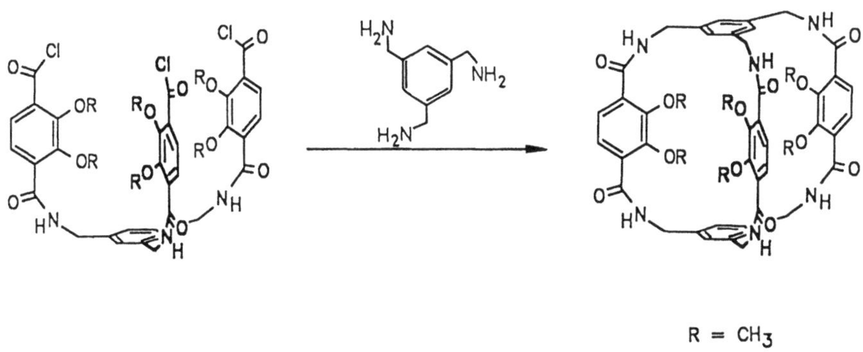

Fig. 5. Preparation of the first triply bridged catechol ligand **4**

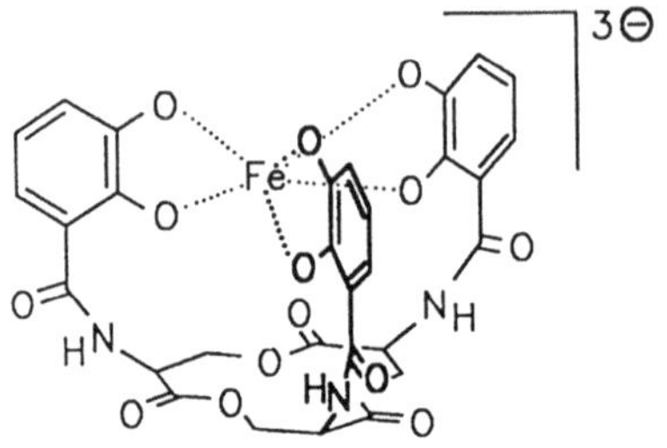

Fig. 6. Structure of Fe(enterobactin)$^{3-}$

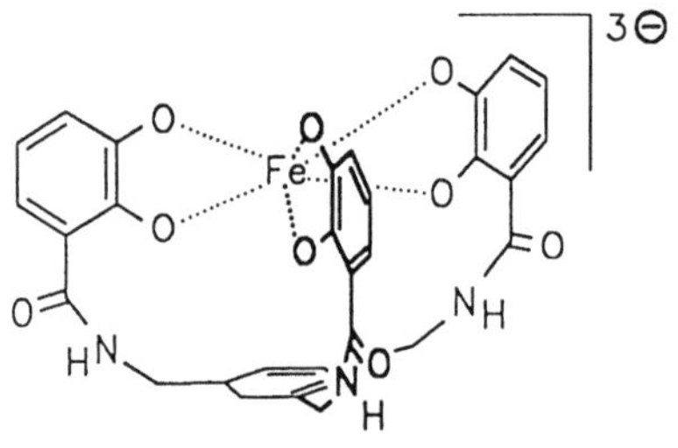

Fig. 7. Structure of Fe(MECAM)$^{3-}$

by the new macrobicyclic ligand **4**. Its extremely high complex constant of $10^{59}\,l\,mol^{-1}$ demonstrates that this ligand is an even stronger Fe^{3+}-binder than enterobactin. The high stability of this complex is demonstrated by its insensitivity to strong acids: Even at pH 2.5 the complex is not destroyed [13]: The macrobicyclic ligand **5** (TRENCAM) synthesized later than **4** shows an unusual trigonal prismatic structure which lends chirality to the complex. All catechol-ligands described so far discriminate against Fe^{2+} in favor of Fe^{3+} whereby the quotient of the complex constant $K_{Fe(III)}/K_{Fe(II)}$ for **5** amounts to 10^{29}.

Complex ligands like **4** or **5** are also useful for the complexation of triply charged cations of main group III of the periodic table of the elements, e.g. the ligand **4** forms a Ga^{3+}-complex.

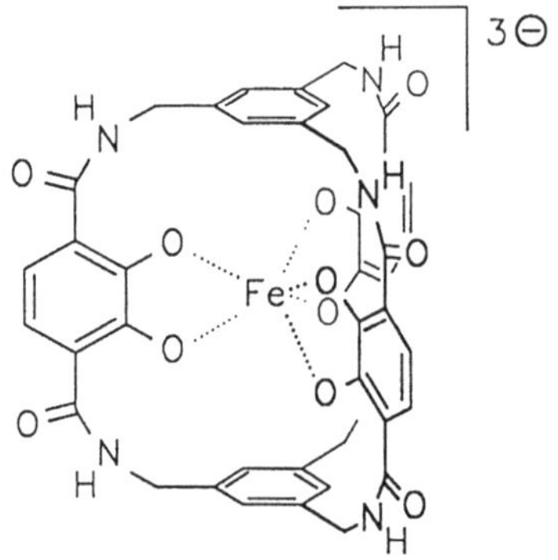

Fig. 8. Structure of the Fe^{3+}-complex of **4**

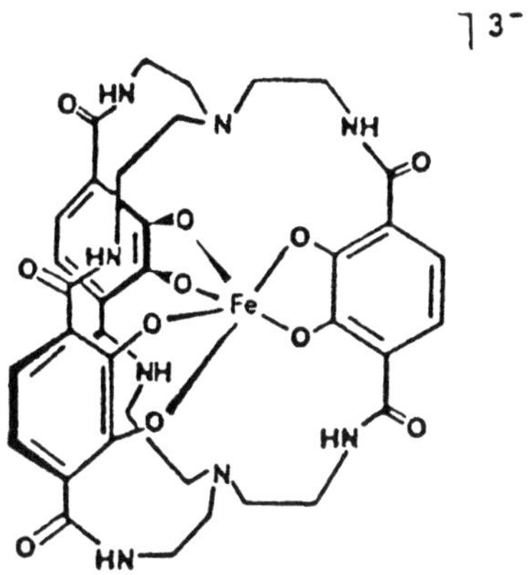

Fig. 9. Fe^{3+} complex of the ligand **5**

3 Selective Molecular Recognition by Macrobicyclic Large Cavities

3.1 Large Molecular Cavities of the Tris-Bipyridine Type

3.1.1 Synthesis

The synthesis of very large cavities of the tris-bipyridine type is usually carried out starting with the corresponding substituted podands through formation of 3 amide bonds in one step with yields ranging between 5 and 20%. As spacer groups hereby 1,3,5-tris-substituted benzene rings serve as well as 1,3,5-triphenylbenzene (substituted in the 4′,4″,4‴-positions or in the 3′,3″,3‴-positions). Unsymmetric, large cavities are also possible as Fig. 10 shows with the example

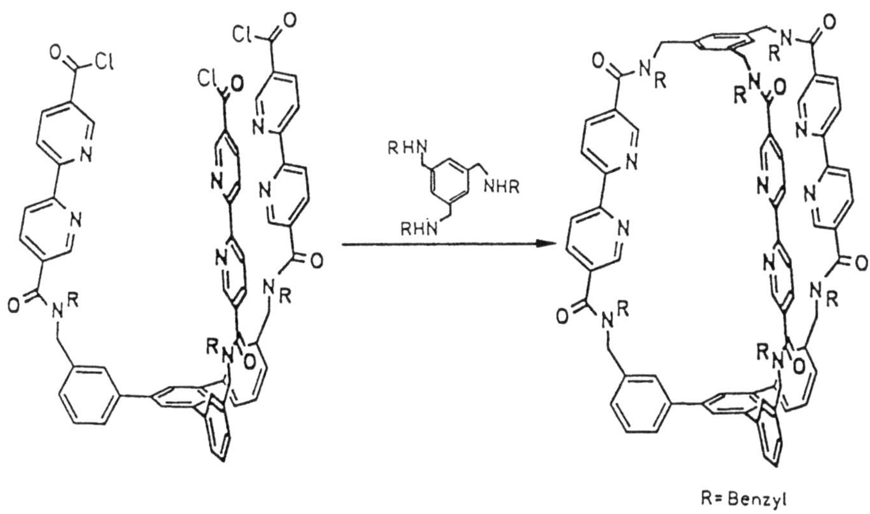

Fig. 10. Synthesis of the macrobicyclic host **6**

of a conical macrobicyclic host skeleton. According to the same principle, the ligands **7** and **8** with even larger "spacer plates" and therefore larger cavities have been prepared [17]: A macrobicyclic cavity with a differing topology but rather similar complexation properties has been synthesized through coupling of acetylenic functions. Two isomers are formed which were separated and investigated for their complexation properties [18].

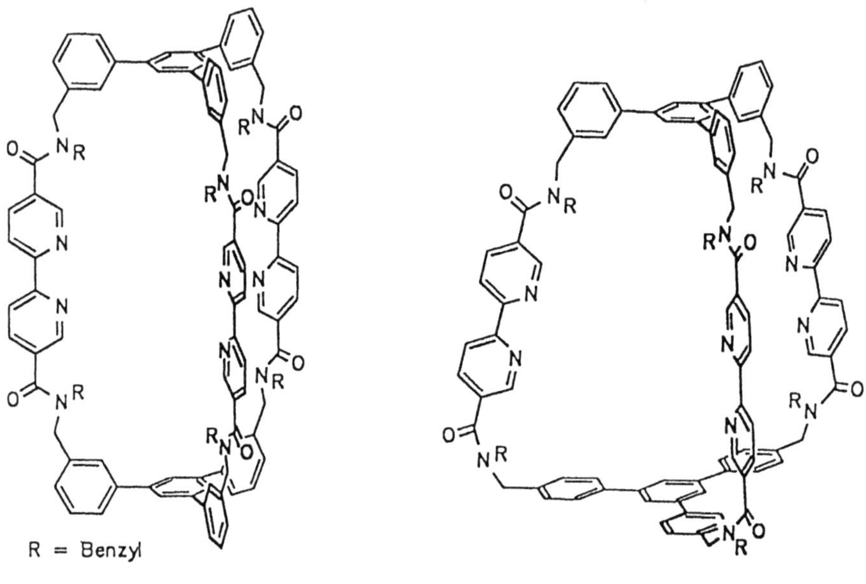

7 8

Fig. 11. Structures of the host skeletons **7** and **8**

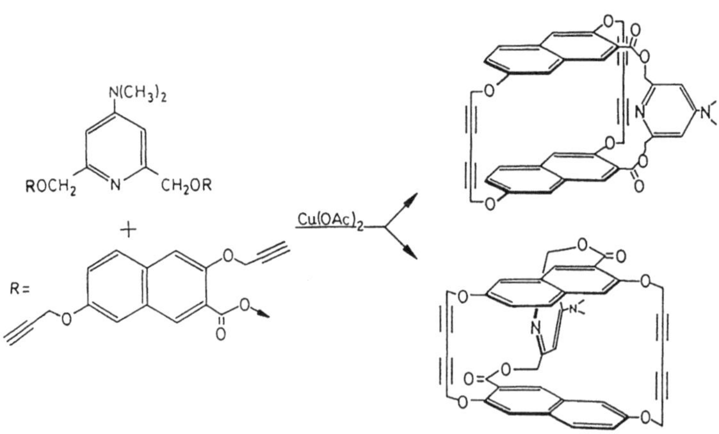

10

9

Fig. 12. Synthesis of the ligands **9** and **10**

3.1.2 Complexation of Phenolic Guest-Molecules

The complexation of organic guest molecules by multiple hydrogen-bond-bridging in nonaqueous media, whereby the solvent does not compete with the host molecule, as well as complex formation with specific hydrogen-bridging is gaining increasing importance [18, 19]. Two possibilities exist: One is to construct the proton donor-functionality inside the host cavity and a proton acceptor at the guest structure [19]. The second possibility is to use proton donating guests capable of forming hydrogen bridges with acceptor host molecules. Nitrogen heterocycles like pyridine or bipyridine are especially suited for the recognition of guest molecules capable of forming hydrogen bridges. The host structures **9** and **10** bind phenolic guest molecules by use of donor-enforced pyridine units (DMAP-moieties) as well as by π–π-interactions, which nevertheless in organic media as $CDCl_3$ are rather small. The isomer **9** binds *p*-nitrophenol with an association constant K_{ass} of 3000 $l\,mol^{-1}$. Phenol itself is loosely bound (K_{ass} = 20 $l\,mol^{-1}$) whereas the isomer **10** for *p*-nitrophenole exhibits an association constant of 13700 $l\,mol^{-1}$. Benzoic acid is only bound by the host isomer **10** (K_{ass} = 5700 $l\,mol^{-1}$). The macrobicyclic ligands **6**, **7** and **8** show a similar complexation behaviour, but in addition the formation of the multiple hydrogen bond bridging in the interior of the host compounds and also the cooperativity of three bipyridine units is observed. Whereas **3** does not show acceptor properties towards organic guest molecules, the host molecules **6** and **7** enclose 1,3,5-trihydroxybenzene (phloroglucinol), 1,2,4-trihydroxybenzene, 2,4,6-trihydroxyacetophenone and 2,4,6-trihydroxybenzaldehyde. For the phloroglucinol complex of **7** an association constant K_{ass} of 11000 $l\,mol^{-1}$ has been determined [20].

The larger host compound **8** only complexes the somewhat more space filling guest molecules 2,4,6-trihydroxyacetophenone and 2,4,6-trihydroxybenz-

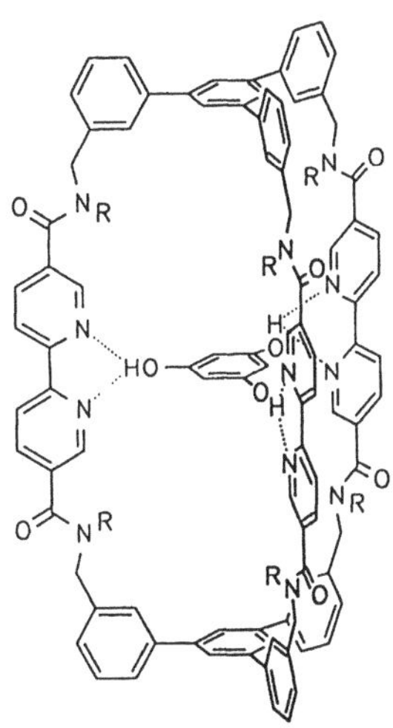

R = Benzyl

Fig. 13. Proposed structure of the complex of the phloroglucinol complex of the host **7**

aldehyde, which hints at a high selectivity of these receptor type molecules with regard to the topology of the guests incorporated. Carboxylic acids like benzene-1,3,5-tricarboxylic acid are not complexed and with more acidic phenols like nitrophloroglucinol the formation of a 1:3-complex is observed by a simple protonation of the bipyridine. This shows that as in the previous example the difference in the pK_A-values of the host and guest should not be too dramatic to give rise to an inclusion of the guest inside the hosts cavity cf. [18a].

3.2 Very Large Molecular Cavities of the Tris-Catechol Type

3.2.1 Synthesis

As has been described in Sect. 2.2.1, triply bridged catechol ligands are synthesized advantageously starting with the corresponding podand intermediates through formation of three amide bonds. Thereby spacer groups of the triphenylbenzene type are added. In this way a row of ligands of the siderophore type with varying cavity sizes is available [21].

3.2.2 Complexation Properties

The host compounds **11**, **12** and **13** were studied with regard to their capability of forming molecular inclusion compounds with triamine guests such as 1,3,5-tris-(aminomethyl)benzene by use of hydrogen bridges. On account of the low solubility of these host compounds, the complexation studies had to be undertaken in DMSO-d_6, a solvent which competes with the hydrogen bridges between host and guest. Only the protonation of the amine guest was observed [21].

The more spacered ligands **11**, **12** and **13** with Fe^{3+} cations lead to tris-catechol complexes but they do not show the high stability of the **4**·Fe^{3+}-complex and which have been shown by UV/VIS-spectroscopy to exist at least partly in an oligomeric or polymeric complex form.

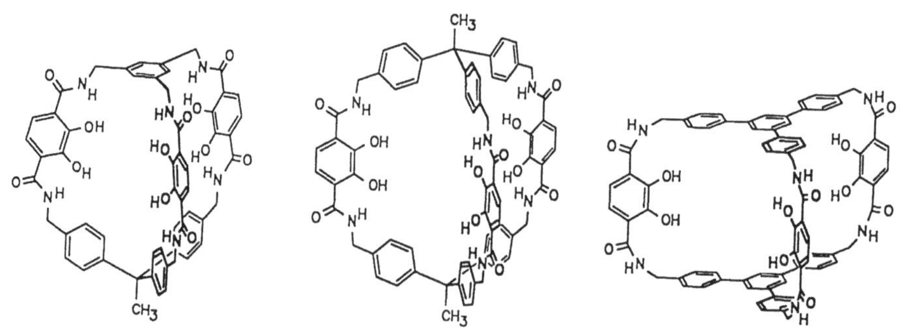

11 12 13

Fig. 14. Structure of the catechol type ligands **11, 12, 13**

3.3 Host Compounds with Large Hydrophobic Cavities – Without Functional Groups Inside the Cavity

3.3.1 Synthesis

In the case of triply bridged host compounds without functional groups inside the cavity, 4,4′-substituted diphenylmethane derivatives are often used as bridge building units, because these form an efficient hydrophobic cleft and in addition are comparatively easily available. Corresponding monocyclic host compounds have been known for years and on account of their good complexation properties towards lipophilic guest molecules meanwhile belong to the classic cases of molecular recognition [22, 23]. The synthesis of macrobicyclic host compounds mostly proceeds by the connection of amide bonds. The host compound **14** for example was tailor-shaped by cyclization of a monocyclic diamine with a dicarboxylic acid dichloride and subsequent reduction of the amide groups with the $BH_3 \cdot$THF-complex [24]. The simpler way to produce such host structures although with lower yields, consists in the reaction of a diamine with a tricarboxylic acid trichloride. The hexaamine **15** was obtained in this way through the reaction of 1,3,5-benzenetricarboxylic acid trichloride with

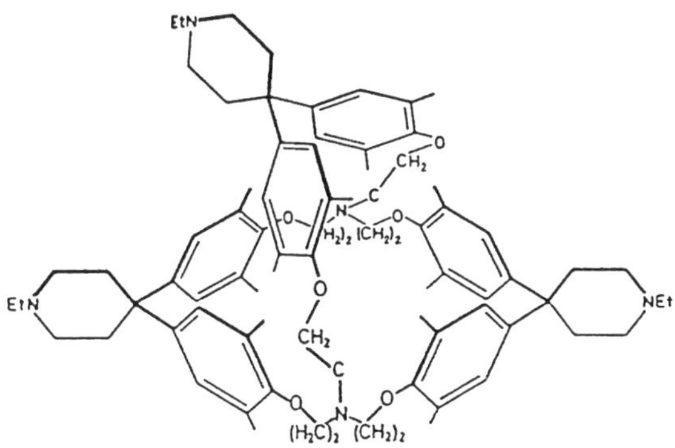

14

Fig. 15. Structure of the macrobicyclic host compound **14**

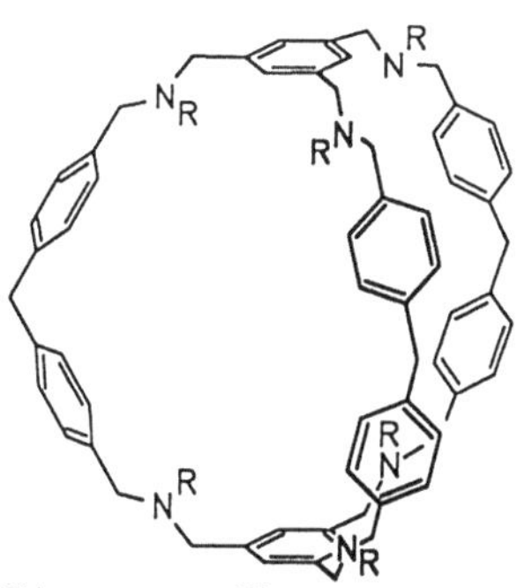

R = CH3 15

Fig. 16. Structure of the macrobicyclic host compound **15**

4,4′-bis(methylaminomethyl)-diphenylmethane and subsequent reduction of the six amide bonds using $BH_3 \cdot THF$ [25].

For complexation in aqueous solutions, mainly *N*-methylated tertiary amines or secondary amines are used because their hydrochlorides are sufficiently water-soluble. For complexation in organic solvents, however, in contrast, one usually uses *N*-benzylated amides (cf. Sect. 3.1), as unsubstituted amides often exhibit low solubility (cf. Sect. 3.2).

Another way to obtain such macrobicyclic host compounds with hydrophobic niches consists of the alkylation of tosylates. Following this synthetic line the host compound **16** is gained by reaction of 4,4′,4″-(*N*-*p*-tosylamino)-1,1,1-triphenylethane with 1,6-dibromohexane in two steps and subsequent splitting of the six tosylate groups. The host molecule **16** exhibits an interesting in/out-isomerism; the complexation properties of both isomers are dramatically different [26]: Whereas the out/out-isomer is capable of complexing adamantane in aqueous solution, the out/in-isomer fails to do this. Both isomers nevertheless are capable of complexing naphthalene derivatives in aqueous solution. Molecular modelling calculations strongly support these experimental findings.

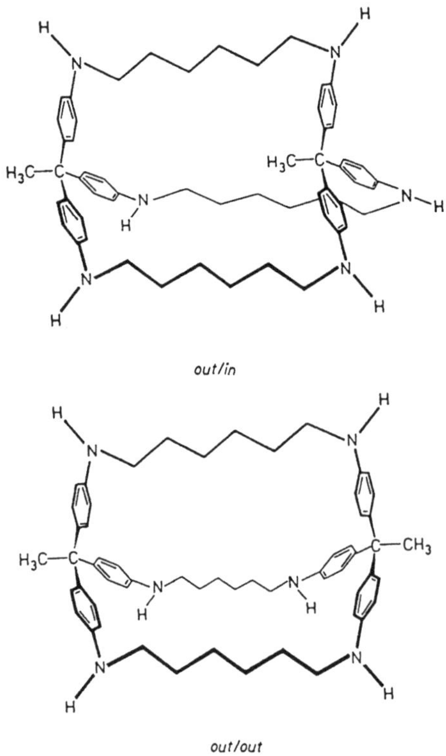

Fig. 17. In/out-isomerism of the large-cavity host molecules **16**

3.3.2 Selective Molecular Recognition of Aromatic Guest Molecules

Host structures of type **14** and **15**, like the mono-cyclic analogous ring compounds, are suited especially for the complexation of aromatic guest molecules in watery phases, as π–π-interactions and especially the "hydrophobic effect" can be applied here. In the case of the triply bridged host analogues the association constants are nevertheless significantly higher. Whereas **14** with pyrene as the guest in D_2O forms a complex with $K_{ass} = 4.1 \times 10^6\ l\,mol^{-1}$, the association constant for the analogues monocyclic host **17** with D_2O is only $K_{ass} = 1.8 \times 10^6\ l\,mol^{-1}$ [24, 27].

For the determination of the association constants in aqueous solution, naphthalene derivatives like 2,6- or 2,7-dihydroxynaphthalene are especially suitable as these are somewhat soluble in water without the addition of host substance. Therefore the ^1H-NMR-shifts of the free guest molecules can serve as references for the determination of the saturation shifts [28]. By determination of association constants at different temperatures the parameters ΔH_{ass} and ΔS_{ass} are obtained which allow a deeper insight into the nature of the host–guest interactions in water solution. Whereas in the case of large host compounds having no tight contact with the guest, entropic factors often prevail because the guest inside the host molecules has a higher mobility than in the solvent itself. In contrast to this, enthalpic factors are dominant if a very tight interaction is possible. The gain in enthalpy has its source on the one hand from Van der Waals interactions of the π-plains, but on the other also because water molecules are thrown out of the host cavities in the course of the encapsulation of the guest. Such water molecules are usually inside the host cavity before the guest enters it. Because of a better solvation of these water molecules in the water

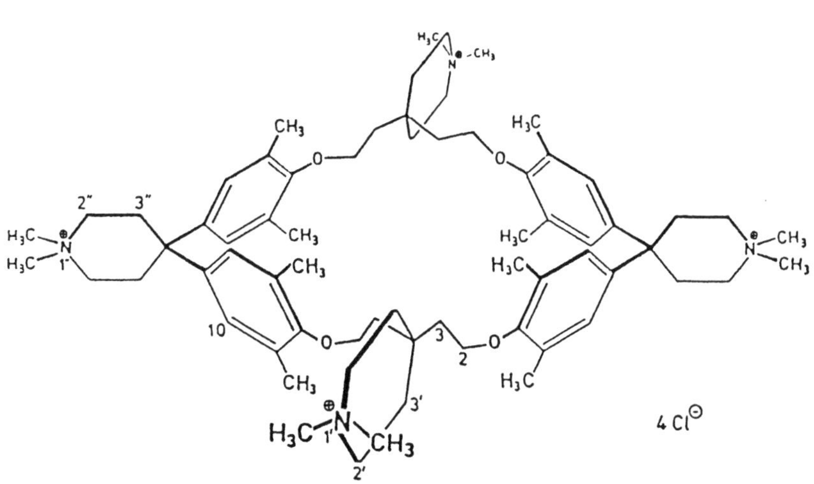

17

Fig. 18. Structure of the monocyclic host molecule **17**

phase than inside the host cavity, an appreciable energy gain results in the course of encapsulating of the guest in the host cavity [29]. In the case of tailor-shaped macrobicyclic host compounds like **14**, **15** and **16** the good complexation properties are therefore mainly due to enthalpic reasons. Unfortunately these processes have, up to now, seldom been studied in a more quantitative way.

Whereas with the host compound **14** a complexation of pyrene and naphthalene derivatives was achieved, the macrobicyclic large cavity molecule **15** allows the selective molecular recognition of isomeric and partially hydrogenated arenes [25]. Aromatic compounds like acenaphthylene, phenanthrene, pyrene and triphenylene are enclosed inside the host cavity, whereby small changes of the topology of the guest molecule lead to a failure of the inclusion. This is the reason why acenaphthene, 3,6-dimethyl phenanthrene and partially hydrogenated pyrenes and triphenylenes are not encapsulated by the host **15**. The out/out-isomer of the host compound **16** has been shown to complex adamantane with a host to guest ratio of 1:1 whilst the out/in-isomer also formed in the synthesis does not complex adamantane. Molecular modelling calculations lead to the result that the out/in-isomer does not possess an energy minimum with adamantane as the guest which can be related to a host–guest inclusion, but on the contrary only an exocyclic niche with a low binding tendency is formed on the outside of the molecule [30].

4 Outlook

The host structures and their properties described in this short progress report show that selective molecular recognition using three-dimensional spherical structures has made new progress. The higher synthetic effort compared to monocyclic host compounds is in many cases more than compensated by new selectivities, higher association constants, higher photostabilities and so on. In addition the synthetic effort can be reduced by use of modular unit construction strategies as described. The next step on the way to abiotic enzyme models will be the recognition of chiral guest compounds as well as investigation of catalytic systems, whereby topics like the molecular self-organisation as well as homogeneous catalyses including transition metal centers lie in the foreground. Apart from improving the selectivities and the catalytic properties it will be necessary to obtain a deeper understanding of the nature of these interactions between hosts and guests, whereby physicochemical methods, quantum chemical calculations, and of course modern synthetic strategies, tactics and methods are required.

5 References

1. a. Weber E (1988) In: Phase transfer catalysis – Properties and applications, Merck-Schuchardt, Darmstadt, p 33

b. Izatt RM, Bradshaw JS, Nielsen SA, Lamb JD, Christensen J (1985) Chem. Rev. 85: 271
2. Dehmlow EV, Dehmlow SS (1980, 1983) Phase transfer catalysis, Verlag Chemie, Weinheim (Monographs in Modern Chemistry, vol 11)
3. a. Rodriguez-Ubis JC, Alpha B, Plancherel D, Lehn JM (1984) Helv. Chim. Acta 67: 2264
b. Alpha B, Anklam E, Deschenaux R, Lehn JM, Pietraskiewicz M (1988) Helv. Chim. Acta 71: 1042
4. Ebmeyer F, Vögtle F (1989) Chem. Ber. 122: 1725
5. Grammenudi S, Vögtle F (1986) Angew. Chem. 98: 1119, (1986) Angew. Chem. Int. Ed. Engl. 25: 1119
6. Belser P, De Cola L, von Zelewsky A: J. Chem. Soc. Chem. Commun. 1988: 1057
7. Vögtle F, Müller WM, Raßhofer W (1979) Isr. J. Chem. 18: 246
8. Photochemical water splitting see e.g.: Vögtle F (1989) Supramolekulare Chemie, Teubner, Stuttgart; Kalyanasundaram K (1982) Coord. Chem. Rev 46: 159
9. Juris A, Balzani V, Barigelletti F, Campagna S, Belser P, von Zelewsky A (1988) Coord. Chem. Rev. 84: 85
10. Alpha B, Balzani V, Lehn JM, Perathoner S, Sabbatini N (1987) Angew. Chem. 99: 1266, (1987) Angew. Chem. Int. Ed. Engl. 26: 1266
11. Dürr H, Zengerle K, Trierweiler H-P (1988) Z. Naturforsch. 43b: 361
12. De Cola L, Barigeletti F, Balzani V, Belser P, von Zelewsky A, Vögtle F, Ebmeyer F, Grammenudi S (1988) J. Am. Chem. Soc. 110: 7210. Barigelletti F, De Cola L, Balzani V, Belzer P, von Zelewsky A, Vögtle F, Ebmeyer F, Grammenudi S (1989) J. Am. Chem. Soc. 111: 4662
13. Kiggen W, Vögtle F (1984) Angew. Chem. 96: 712, (1984) Angew. Chem. Int. Ed. Engl. 23: 714; Kiggen W, Vögtle F, Franken S, Puff H (1986) Tetrahedron 42: 1859
14. a. McMurry TJ, Rodgers SJ, Raymond KN (1987) J. Am. Chem. Soc. 109: 3451
b. McMurry TJ, Hosseini MW, Garrett TM, Hahn FE, Reyes ZE, Raymond KN (1987) J. Am. Chem. Soc. 109: 7196
15. Peter-Katalinić J, Ebmeyer F, Seel C, Vögtle F (1989) Chem. Ber. 122: 2391
16. Harris WR, Weitl FL, Raymond KN: J. Chem. Soc. Chem. Commun. 1979: 177
17. Ebmeyer F, Vögtle F (1989) Angew. Chem. 101: 95, (1989) Angew. Chem. Int. Ed. Engl. 28: 79
18. a. Sheridan RE, Whitlock HW (1986) J. Am. Chem. Soc. 108: 7120
b. Sheridan RE, Whitlock HW (1988) J. Am. Chem. Soc. 110: 4071
19. Rebek J (1988) Topics Curr. Chem. 149: 189
20. See [17], cf. Watson WH, Vögtle F, Müller WM (1988) J. Incl. Phenom. 6: 491
21. Stutte P, Kiggen W, Vögtle F (1987) Tetrahedron 43: 2065
22. Odashima K, Itai A, Iitaka Y, Koga K (1980) J. Am. Chem. Soc. 102: 2504
23. Overview: Diederich F (1988) Angew. Chem. 100: 372
24. Diederich F, Dick K (1984) Angew. Chem. 96: 789
25. Vögtle F, Müller WM, Werner U, Losensky H-W (1987) Angew. Chem. 99: 930, (1987) Angew. Chem. Int. Ed. Engl. 26: 901; Wallon A, Peter-Katalinić J, Werner U, Müller WM, Vögtle F (1990) Chem. Ber. 123 in press
26. Franke J, Vögtle F (1985) Angew. Chem. 97: 224, (1985) Angew. Chem. Int. Ed. Engl. 24: 219
27. Diederich F, Dick K (1984) J. Am. Chem. Soc. 106: 8024
28. a. Diederich F, Griebel D (1984) J. Am. Chem. Soc. 106: 8037
b. Wilcox CS, Cowart MD (1986) Tetrahedron Lett. 27: 5563
29. Diederich F (personal commun.) Sept. 1988; cf. Schneider HJ, Kramer R, Simova S, Schneider U (1988) J. Am. Chem. Soc. 110: 6442
30. Franke J, Vögtle F (unpublished); cf. F. Vögtle et al.: Stereochemie in Stereobildern. VCH-Verlag, Weinheim 1987

Functionalization of Crown Ethers and Calixarenes: New Applications as Ligands, Carriers, and Host Molecules

Seiji Shinkai
Department of Organic Synthesis, Faculty of Engineering, Kyushu University, Fukuoka 812, Japan

Bioorganic Chemistry Frontiers, Vol. 1

1 Introduction

Since the unexpected discovery of dibenzo-18-crown-6 by Pedersen in 1967, the chemistry of the crown ether family has rapidly become established as a new field. The synthesis of new crown compounds and the characterization of their metal complexes became the main objects of investigation. Similar progress has been seen for calixarene chemistry. Although synthesis was established quite early on, it drew little attention from chemists. Later, Gutsche and co-workers established the convenient one-step synthesis of calixarenes. After that, the chemistry of a calixarene family has been expanding rapidly. On the other hand, the functional facet of crown ethers and calixarenes has been left less exploited for a long time. Looking over the role of natural ionophores, for example, we notice that they are designed so that they can satisfy the functional requirements of nature and "work" efficiently in a biological system. This suggests that the next investigation efforts should be devoted to functionalization of these supramolecules, which would serve eventually as a crosslink between the synthesis and the functional facet. In this chapter, we review strategies for functionalization of these molecules so that they can "work" in artificially-designed, biomimetic systems.

2 Switch-Functionalized Crown Ethers

Certain natural ionophores such as monensin and nigercin not only recognize the metal cation but also "work" only when they are excited by the stimulus from certain circumstances. One may summarize the essential function of these natural ionophores as follows: they respond to the specific metal cations under the specific stimuli. The basic idea of "switched-on" crown ethers originates from the responsive action of these natural ionophores: that is, this functionality is attained by the combination of an antenna molecule which acts as a stimulus-responsive trigger and a functional group (crown ether) which causes a subsequent event within the same molecule.

More concrete ideas are found in biological systems. Iron is transported across a biomembrane by ionophores called "siderophores": the stability constants are so large (K_s = approx. 10^{20}–$10^{50}\ M^{-1}$) that the rate of decomplexation by which the iron is released from the membrane phase to the receiving aqueous phase is limited by the transport system [1]. Thus, the K_s should be reduced in order to facilitate the ion-release. The strategy employed by the biological system is either to reduce Fe^{3+} to Fe^{2+} or to decompose the siderophores by enzymes. The sodium transport method employed by monensin seems more promising for designing "switched-on" systems: it utilizes a pH-induced dissociation of the terminal carboxylate group to attain conformational change between the cyclic (high K_s) and the noncyclic form (low K_s), which leads to the efficient dynamic transformation between the ion-complexation site and the ion-release site of the membrane [2]. This means that, in an artificial ion-transport system using crown ethers, if the stability constants can be changed in response to the environmental conditions, a variety of functions related to crown ethers can be controlled by an on-off switch [3].

2.1 pH-Responsive Crown Ethers

The oldest example for the switch-functionalized systems is the class of pH-responsive crown ethers. These seem to be direct mimics of natural polyether antibiotics such as monensin, nigericin, lasalocid, etc. These polyether antibiotics have a hydroxyl on one end and a carboxyl group on the other end of the polyether chain in order to effect the cyclic–noncyclic interconversion through the formation and scission of the hydrogen-bond. Typical examples for noncyclic analogs are compounds (**1**)–(**4**) which have these two functional groups at the chain ends and mimic the ion transport properties of natural polyether antibiotics [4–6]. In fact, it is known that these ionophores can carry metal cations across a membrane from a basic aqueous phase to an acidic aqueous phase as do the natural antibiotics. (**4**) is known to show the high affinity for alkaline earth metal cations (Ca^{2+}, Ba^{2+}) over alkali metal cations [6].

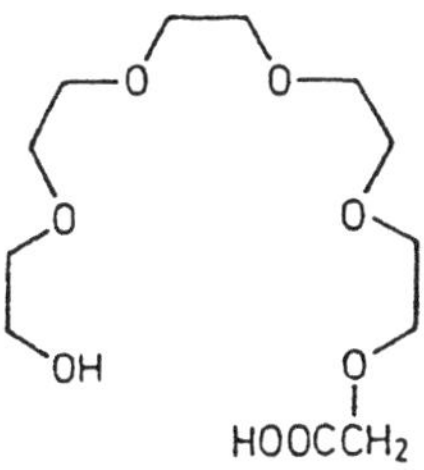

(1)

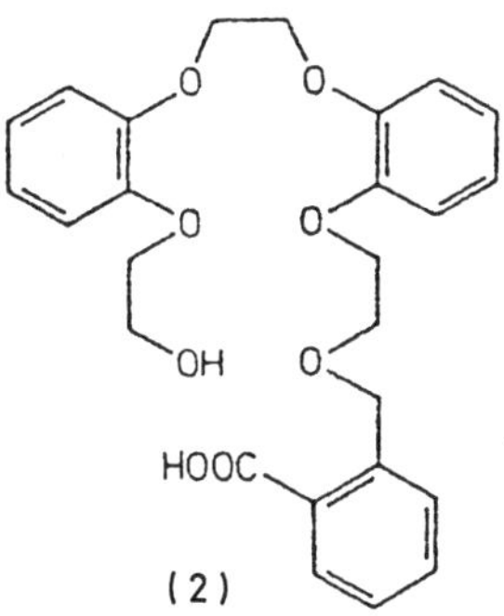

(2)

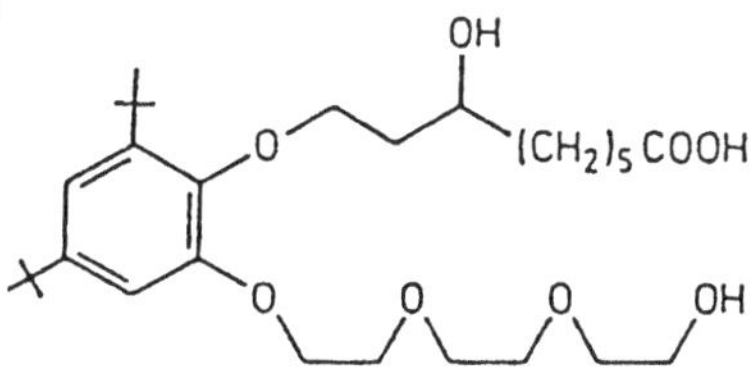

(3)

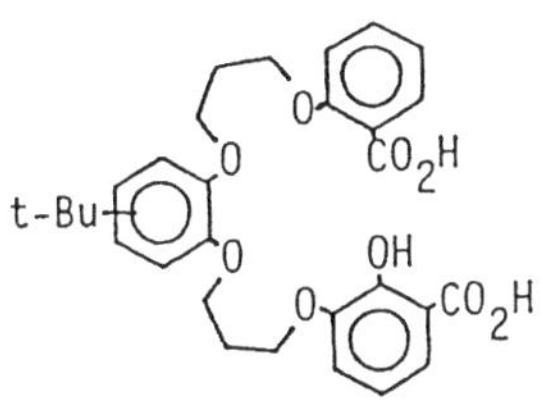

(4)

The more direct method of lowering the stability constant by the pH change is to protonate the ring nitrogen. Matsushima et al. [7] demonstrated that monoazacrown ether derivatives (**5**) act as good carriers for active transport of alkali metal cations from the basic IN aqueous phase to the acidic OUT aqueous phase. The transport ability is well correlated with the complexation ability, indicating that the ion-uptake from the basic IN aqueous phase is rate-limited while the ion-release to the acidic OUT aqueous phase is relatively fast. This means that facile decomplexation takes place by protonation of the ring nitrogen. Another example related to the nitrogen protonation is a "tail-biting" crown ether (**6**) [8,9]. (**6**) is designed so that the intramolecular complexation between the crown and the ammonium tail can occur in the acidic pH range and enforce the release of the metal cation from the crown-metal complexes. As expected, (**6**) acts as an ion carrier for active transport from the basic to the acidic aqueous phase.

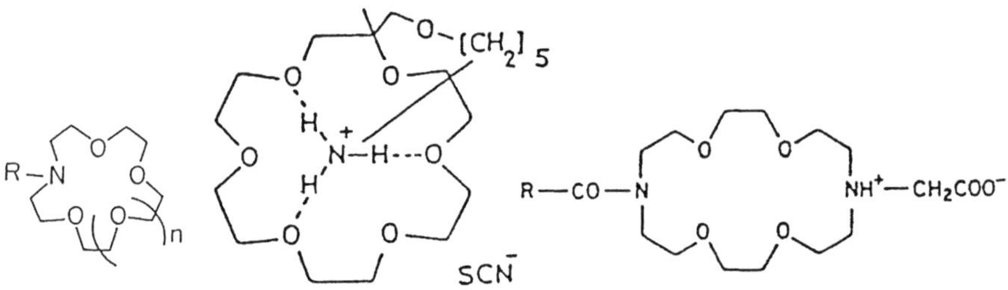

(5) (6) (7)

In (**5**) the ion-release from the membrane is effected by the protonation of the ring nitrogen, so that some fraction of the protonated azacrown ethers may leak into the acidic aqueous phase. In order to avoid this disadvantage Shinkai et al. [10] designed (**7**) (pK_a 3.45, 6.99): that is, the ion-release from the membrane is effected by protonation of the ring notrogen but the product is the lipophilic zwitterionic species. Hence, (**7**) can release the ion into the neutral pH solution and the resultant zwitterionic (**7**) is still retained in the membrane phase. In fact, it was demonstrated that (**7**) does not leak into the neutral aqueous solution and efficiently transports alkali and alkaline earth metal cations in an active transport manner from the basic aqueous phase to the neutral aqueous phase.

It is well-known that metal-proton coupled transport can be mediated more efficiently by anion-capped crown ethers such as (**8**)–(**10**) [11–14]. These crown derivatives exhibit not only the excellent metal affinity but the high metal selectivity due to the crown ether ring while the transport ability is controlled by the pH-responsive carboxylate cap placed exactly on the ring. When the anionic cap acts cooperatively with the crown ether ring, they serve as carriers for active transport of alkali and alkaline earth metal cations (Ca^{2+} and Ba^{2+}) at a speed comparable to certain antibiotics [11].

2.2 Redox-Switched Crown Ethers

Crown ethers which have a redox-active group within the same molecule are classified as redox-switched crown ethers. One can expect that i) the ion-binding ability of the crown ether site can be controlled by the redox state of the prosthetic redox-active site, or on the contrary ii) the redox potential of the redox-active site can be controlled by the metal binding to the prosthetic crown

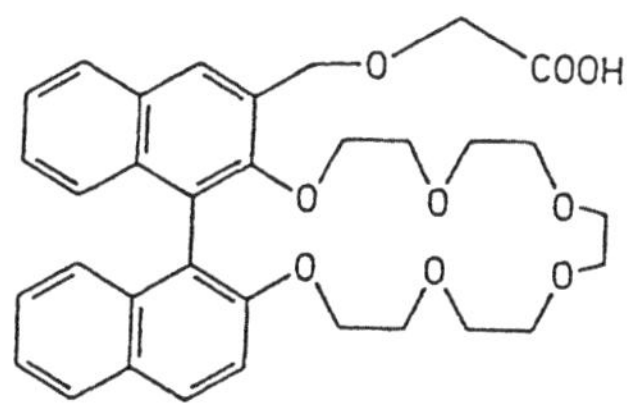

(8)

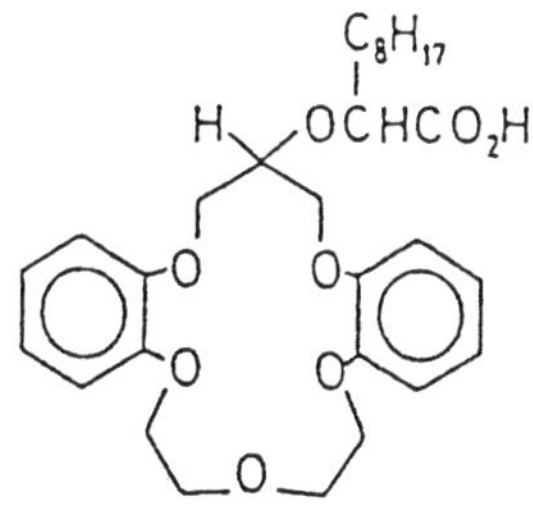

(9)

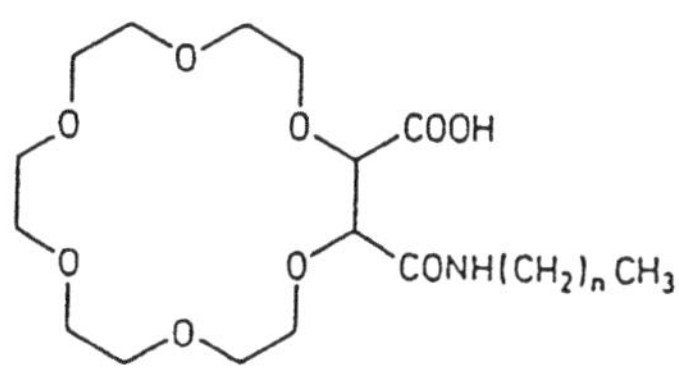

(10)

ether site. When these two sites influence each other, one may consider that they "conjugate" each other. These systems serve as interesting mimics of conjugate events occurring in the nerve cell.

Crown ethers (**11**) having the quinone moiety as a redox-functional group were first synthesized by Misumi et al. [15, 16]. They found that the stability constant for (**11**: n = 2)$_{ox}$ is relatively small ($\log K_s = 1.8$ for Na^+) but that for (**11**: n = 2)$_{red}$ ($\log K_s = 2.39$) is comparable with that for (**12**: n = 2) ($\log K_s = 2.51$). This indicates that the intraannular carbonyl oxygen in (**11**)$_{ox}$ is not connected with the binding of metal cations, whereas the hydroxylic oxygen in (**11**)$_{red}$ shows a positive influence on the binding of metal cations. A more detailed investigation on the redox properties was reported by Wolf and Cooper [17] on the 3,5-dimethyl derivative of (**11**: n = 2). In particular, cyclic voltammetric studies provided strong evidence for the desired coupling between the complexation and the redox state: the presence of alkali metal salts makes the quinone easier to reduce. Replacement of 0.1 M $Et_4N^+ClO_4^-$ as supporting electrolyte with 0.1 M $M^+ClO_4^-$ ($M^+ = Li^+$, Na^+) or ($M^+ = K^+$) changes (**11**: n = 2)$_{ox}$/(**11**$^-$: n = 2) formal potential from -0.60 V to -0.55, -0.48, and -0.47 V, respectively, shifts of $+50$, $+120$, and $+130$ mV. These shifts reflect metal cation binding by the crown group and indicate that redox reactions and cation binding can influence each other. Similar quinone-containing crown ethers (**13**)–(**16**) have been synthesized by several groups and are shown to conjugate with the metal binding site [18–21].

The more direct change induced by the redox reaction would be reversible bond formation and bond scission leading to the cyclic–noncyclic interconversion. The redox reaction of a thiol-disulfide couple would be the most suitable candidate for this. Shinkai et al. [22, 23] synthesized new "redox-switched" crown ethers bearing a disulfide bond in the ring (**17**)$_{ox}$ and a dithiol group at its

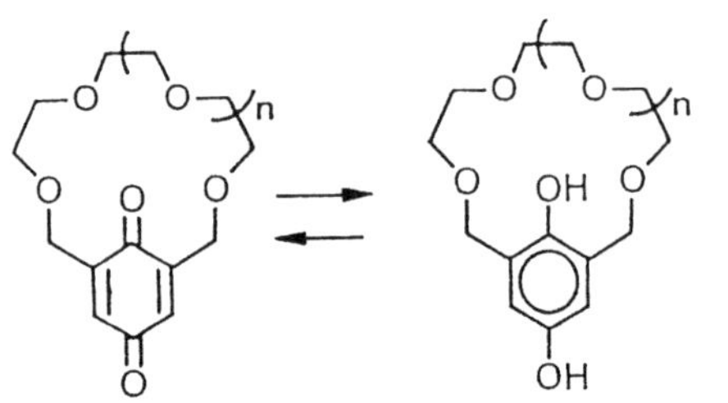

(11)$_{ox}$ (11)$_{red}$

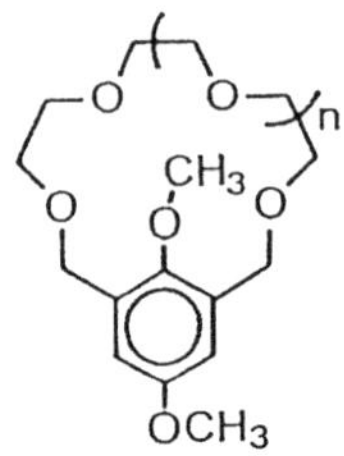

(12)

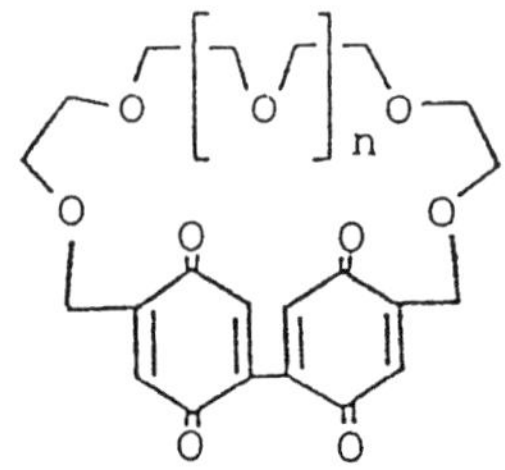

(13)

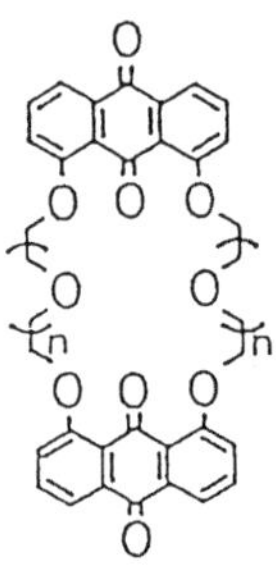

(14)

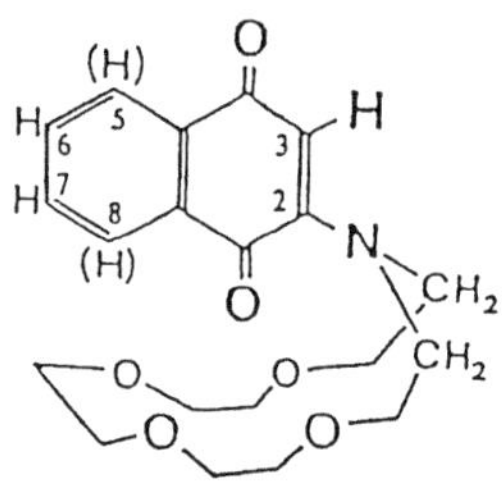

(15)

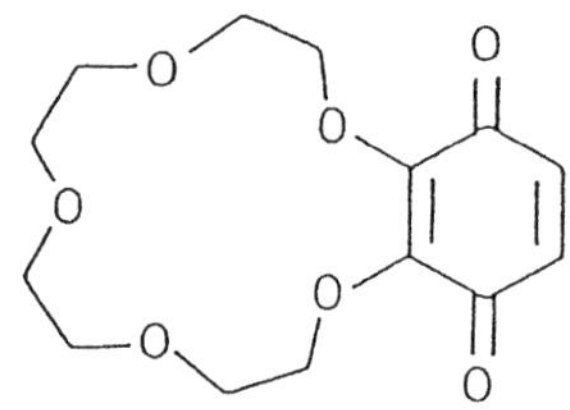

(16)

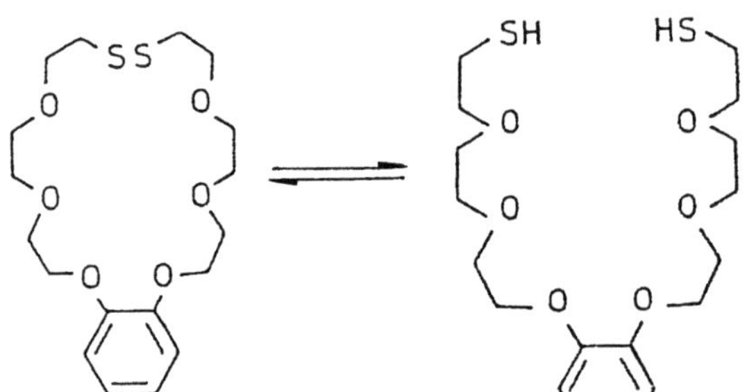

(17)$_{ox}$ (17)$_{red}$

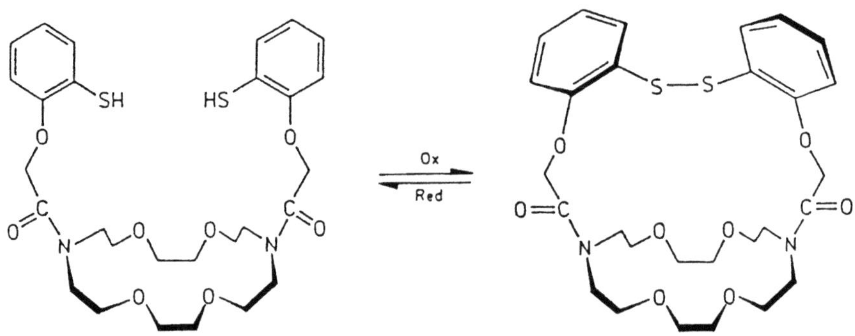

(18)$_{red}$ (18)$_{ox}$

chain ends $(\mathbf{17})_{red}$. A similar approach was also reported by Raban et al. [24]. In ion transport across a liquid membrane, $(\mathbf{17})_{ox}$ carried Cs^+ 6.2 times faster than $(\mathbf{17})_{red}$. It thus became possible to regulate the rate of Cs^+ transport by the redox reaction between $(\mathbf{17})_{red}$ and $(\mathbf{17})_{ox}$ in the membrane phase. One may regard that this is a redox-induced interconversion between coronand and podand. Compound (**18**) was synthesized to demonstrate an interconversion between crown $(\mathbf{18})_{red}$ and cryptand $(\mathbf{18})_{ox}$ [25]. Na^+ $(\mathbf{18})_{red}$ and $(\mathbf{18})_{ox}$ showed a similar ion affinity, but $(\mathbf{18})_{ox}$ bound K^+, Rb^+, and Cs^+ more strongly than $(\mathbf{18})_{red}$ because of the coordination of the cap oxygens to the complexed metal cations. The difference was rationalized by the fact that K^+ (Rb^+ and Cs^+ also) which perches shallowly on the crown ring can interact with the cap oxygens whereas Na^+ which nests deeply in it is too far away to interact with them.

Another method of controlling the binding constants by a thiol redox-switch is an interconversion between monocrown and biscrown. This idea can be realized by using 4′-mercaptomethylbenzo-15-crown-5 [26]. That is, only the oxidized form is capable of forming an intramolecular 1:2 sandwich complex with a metal cation and therefore shows the selectivity toward large alkali metal cations [26]. It was shown that in K^+ transport, the rate is efficiently accelerated when the reduced form is oxidized to the disulfide biscrown ether by iodine added to the membrane phase [26].

The concept of the redox-switch described above would be further extended to the electrochemical switch because the redox-active groups are mostly responsive to the electrochemical signals. Basically, the redox-switched crown ethers become the latent candidates for this class of crown ethers. The electrochemical switch is sometimes superior to the treatment with redox reagents, because one can keep the system clean as long as the electrochemical reaction is reversible. From a biomimetic view point, to give an electrochemical switch to a crown family is also fascinating because the coupling of the electrochemical switch with transport phenomena mimics many biological events which occur in nerve cells. The investigations in this field are reviewed by Gokel in another chapter of this book.

2.3 Photoresponsive Crown Ethers

Photoresponsive systems are seen ubiquitously in nature, and light is coupled with the subsequent life processes. In these systems, a photoantenna to capture a photon is skillfully combined with a functional group to mediate some subsequent events. It is an important fact that these events are frequently linked with photoinduced structural changes of photoantennas. This suggests that in an artificial photoresponsive system chemical substances which exhibit photoinduced structural changes may serve as potential candidates for the photoantennas. In the past, photochemical reactions such as (*E*)–(*Z*) isomerism of azobenzene, dimerization of anthracene, spiropyran-merocyanine interconversion, etc. have been used as practical photoantennas. One can expect that if one of these photoantennas is skillfully combined with a crown ether, many physical and chemical functions of a crown ether family would be controlled by an on–off "light" switch. This is the basic idea for designing photoresponsive crown ethers.

2.3.1 Azobenzene-Bridged Crown Ethers

It is well-known that (*E*)-azobenzene is isomerized to (*Z*)-azobenzene by UV light and the (*Z*)-isomer is isomerized back to the (*E*)-isomer by visible light. Compound (**19**) has a photofunctional azobenzene cap on an N_2O_4 crown ring, so that one can expect that the conformational change in the crown ring would occur in response to the photoinduced configurational change in the azobenzene cap. As expected, (*E*)-(**19**) having the (*E*)-azobenzene cap selectively binds Na^+ while (*Z*)-(**19**) produced by photoisomerization binds K^+ more strongly [26, 27]. This finding suggests that the N_2O_4 ring is apparently expanded by the photoinduced (*E*)-to-(*Z*) isomerization. It was confirmed by X-ray crystallographic studies that in (*E*)-(**20**) the N_2O_4 crown has two anti C–C bonds, resulting in an oval-shaped crown ring [29]. Conceivably, this is the reason (*E*)-(**19**) favors small Na^+ rather than K^+.

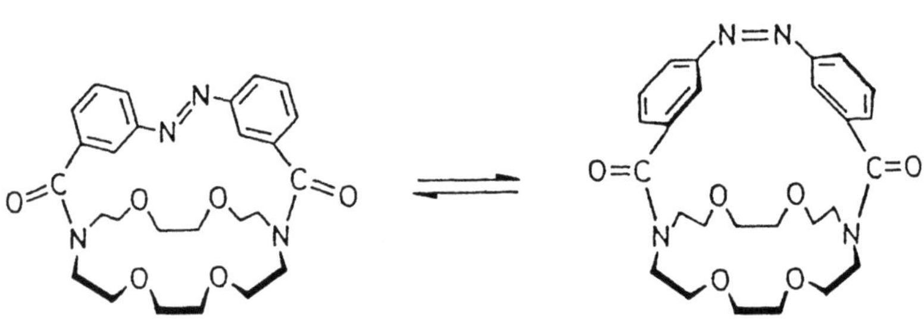

(*E*)-(**19**) (*Z*)-(**19**)

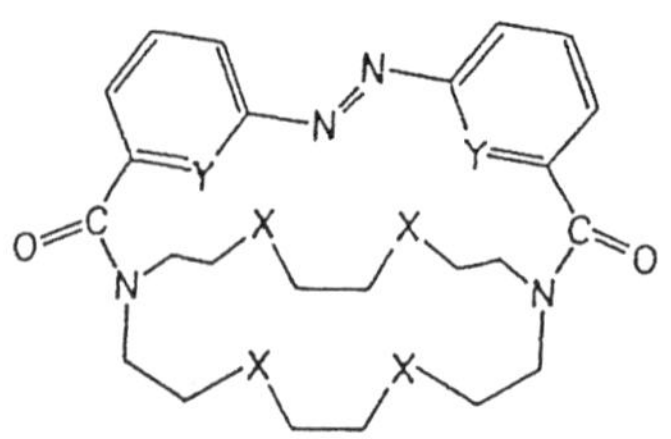

(*E*)-(**20**) : X=O, Y=N
(*E*)-(**21**) : X=S, Y=CH
(*E*)-(**22**) : X=S, Y=N

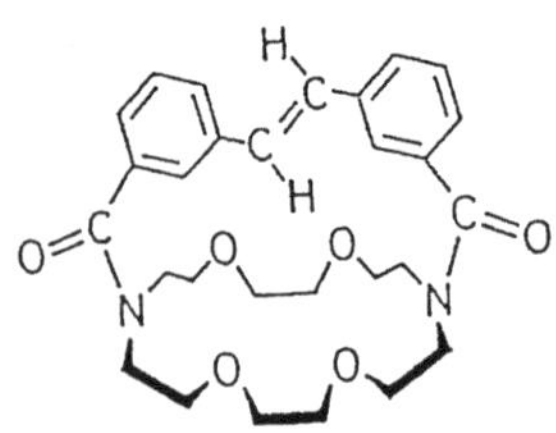

(*E*)-(**23**)

(*E*)-(**20**), in which the azobenzene cap in (*E*)-(**19**) is replaced by a 2,2′-azopyridine, is a cryptand analogue and strongly binds certain heavy metal cations (particularly, Cu^{2+}) [30]. On the other hand, (*Z*)-(**20**) shows no binding ability for Cu^{2+}. These results indicate that pyridine nitrogens of (*E*)-(**20**) are directed toward the crown ether plane so as to cooperatively bind the metal cation, whereas those of (*Z*)-(**20**) have no such coordinating ability due to the distorted configuration. The lack of the affinity for heavy metal cations indicates the importance of the structure of the pyridine cap. The photocontrol of heavy metal binding is also effected by using N_2S_4-containing (**21**) and (**22**) [31]. Another interesting example is a stilbene-capped crown ether (**23**) [32]. Stilbene shows not only (*E*)–(*Z*) isomerism but also fluorescence emission. It was found that (**23**) has as ion selectivity similar to (**19**) and the fluorescence intensity of the (*E*)-isomer is selectively quenched by NaI among alkali iodides tested. This suggests the potential application of (**23**) as a metal-selective fluorescence probe.

2.3.2 Cyclophane-Type Crown Ethers

Photodimerization of anthracene is also usable as a photochemical switch for photoresponsive crown ethers. An early example is (**24**) in which two anthracenes are linked by a polyoxyethylene chain. Photoirradiation of (**24**) in the presence of Li^+ gives the photocyclo-isomer (**25**) [33, 34]. It is known that (**25**) is

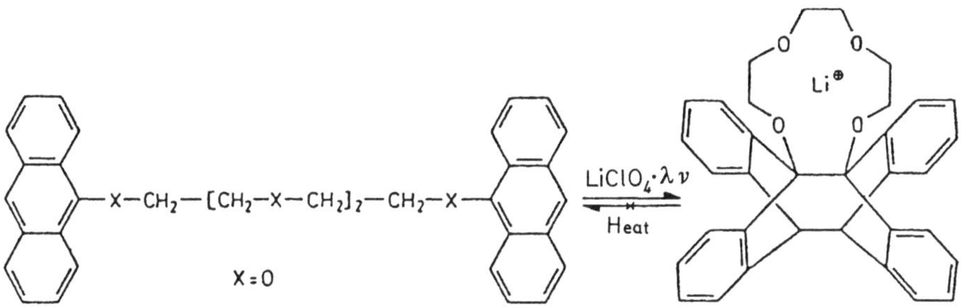

(24) (25)

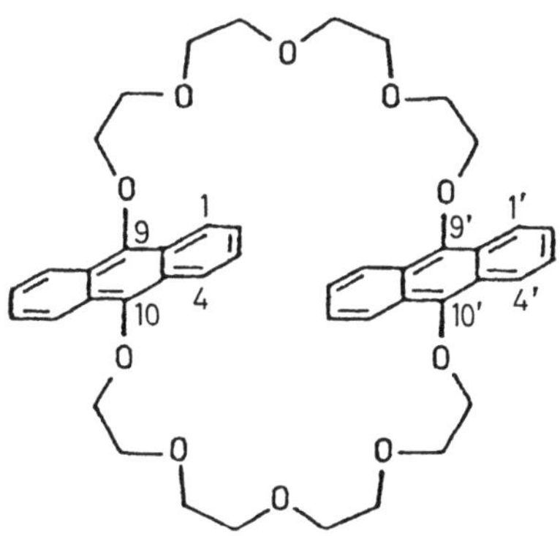

(26)

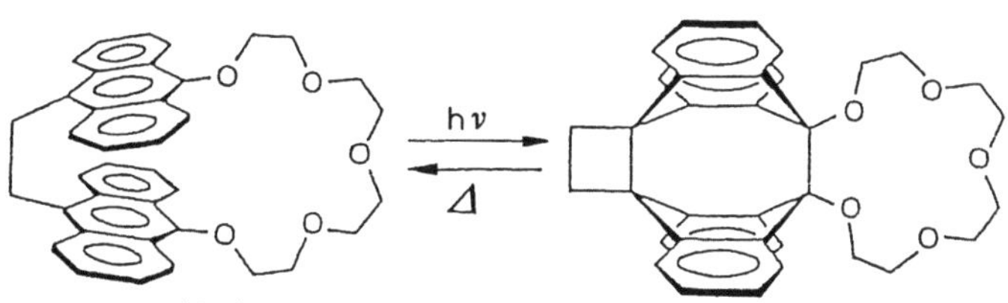

(27) (28)

fairly stable with Li^+ but readily reverted to the open form (**24**) when Li^+ is removed from the ring. In this system, however, one must take into account that intermolecular dimerization may take place competitively with proposed intramolecular dimerization. To rule out this possibility, (**26**) was synthesized in which two anthracenes are included in a ring [35]. It was found that intramolecular photodimerization proceeds rapidly in the presence of Na^+ as the template metal cation. Yamashita et al. [36] also synthesized (**27**) in which intermolecular photodimerization of anthracene is completely suppressed. The photochemically-produced cyclic form (**28**) showed an excellent Na^+ selectivity.

Although the reversibility in photodimerization of anthracene is excellent, the synthesis of anthracene-containing crown ethers is rather troublesome and therefore the application is limited. It thus seems to us that the synthesis of azobenzene-containing alternates would be much easier. Shinkai et al. [37] synthesized azobenzenophane-type crown ethers (**29**) (n = 1, 2, 3) with the 4,4′-positions of azobenzene linked by a polyoxyethylene chain. Examination of the CPK model reveals that the polyoxyethylene chain of the (*E*)-homologs with 6 to 10 ethylene number is extended almost lineraly on the azobenzene plane. Furthermore, the (*E*)–(*Z*) photoisomerization occurs reversibly, indicating that the photoisomerization process is not subject to steric strain. Two-phase solvent extraction indicated that the (*E*)-homologs totally lack the metal affinity while the (*Z*)-homologs bind alkali metal cations, the selectivity for which depends on the polyoxyethylene chain length. The result implies that the polyoxyethylene chain in the (*Z*)-homologs is sterically relaxed enough to form a crown-like loop.

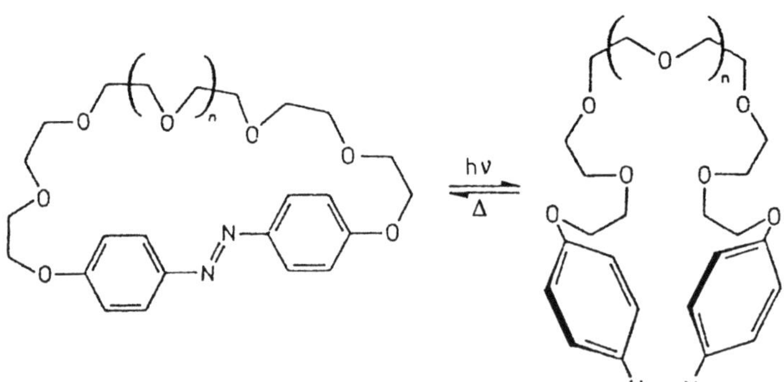

(*E*)-(**29**) (*Z*)-(**29**)

Crown ethers of type (**30**), bearing an intraannular substituent X, bind metal cations to different extents, depending on the nature of X. When X has no metal-coordination ability, it simply provides steric hindrance and decreases the binding constant. In contrast, when it has metal-coordination ability (X = OH, COOH, NH_2, etc.), it increases the binding constant because of the cooperative action of X and the crown ring. Thus, the molecular design of the compound in which X is switched photochemically between (**30**) and (**31**) is very interesting.

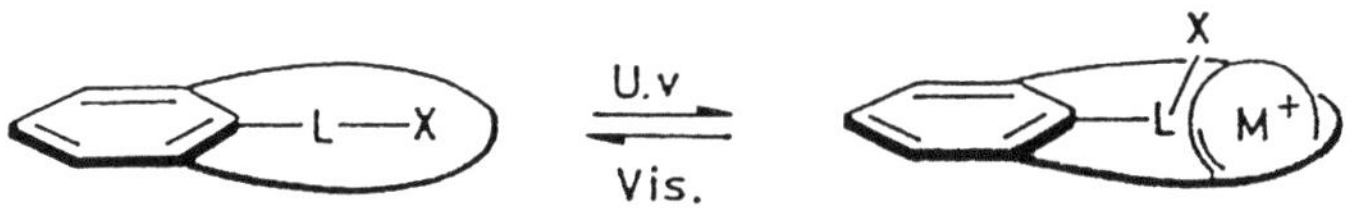

(30) (31)

The azo substitutent in (*E*)-(**32**) simply provides steric hindrance while that in (*Z*)-(**32**) can coordinate to the metal cation complexed in the photochemically-opened crown cavity [38]. As expected, the binding constants (K_s) of the (*Z*)-isomers were estimated to be $10^{4.07}$–$10^{4.81}$ M^{-1} (*o*-dichlorobenzene–methanol = 5:1 v/v), which are comparable with the K_s for regular crown ethers. The detailed spectroscopic examination showed that the azo group in (*Z*)-(**32**) coordinates to the complexed metal cation to enhance the K_s [38]. In contrast, the (*E*)-isomers showed no affinity for Na^+ and only weakly interacted with K^+. This indicates that the (*E*)-azo group simply provides steric hindrance with the approach of guest metals.

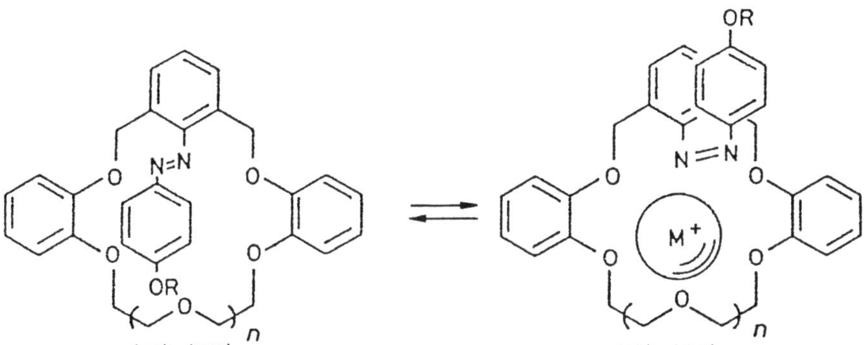

(*E*)-(32) (*Z*)-(32)

These examples consistently indicate that the metal affinity and metal selectivity can be controlled by light which changes the cavity shape by means of the geometrical change in the photofunctional group.

2.3.3 Bis(Crown Ethers)

It has been established that alkali metal cations which exactly fit the size of the crown ether form a 1:1 complex, whereas those which have larger cation radii form a 1:2 sandwich complex. It is known that this phenomenon is the main reason the metal selectivity on the basis of the "hole-size selectivity" is lowered. From a different viewpoint, however, this phenomenon is useful for catching metal cations greater than the crown cavity. This view was clearly substantiated with bis(crown ethers). For instance, Kimura et al. [39] reported that the maleate diester of monobenzo-15-crown-5 ((*Z*)-form) extracts K^+ from the aqueous phase 14 times more efficiently than the fumarate counterpart ((*E*)-form). The difference stems from the formation of the intramolecular 1:2

complex with the (*Z*)-form. If the C=C double bond is replaced by the photofunctional azo-linkage, the resultant bis(crown ethers) would exhibit interesting photoresponsive behaviors. That is, the crown rings in the (*E*)-form would act independently while those in the (*Z*)-form would act cooperatively to form intramolecular sandwich complexes. The basic idea described in this section is related to the photoinduced change in the spacial distance between two crown rings.

Shinkai et al. [40–43] synthesized a series of azobis(benzocrown ethers) called "butterfly crown ethers" such as (**33**) and (**34**). It was found that the content of the (*Z*)-forms at the photostationary state is remarkably increased with increasing concentration of alkali metal cations which can interact with two crown rings in a 1:2 sandwich manner. Similarly, thermal (*Z*)-to-(*E*) isomerization was efficiently suppressed by these metal cations. These findings are clearly ascribable to the crosslink effect of the metal cations with the two crowns in the (*Z*)-forms. In two-phase solvent extraction, the (*Z*)-forms extracted alkali metal cations with large ion radii more efficiently than the corresponding (*E*)-forms. In particular, the photoirradiation effect on (**33**) is quite remarkable: for example, (*E*)-(**33**: n = 2) extracts Na^+ 5.6 times more efficiently than (*Z*)-(**33**: n = 2), whereas (*Z*)-(**33**: n = 2) extracts K^+ 42.5 times more efficiently than (*E*)-(**33**: n = 2) [41]. Therefore, both the metal affinity and the metal selectivity of azobis(benzocrown ethers) change in response to photoirradiation. This novel finding is applicable to light-driven ion transport across membranes (vide ante).

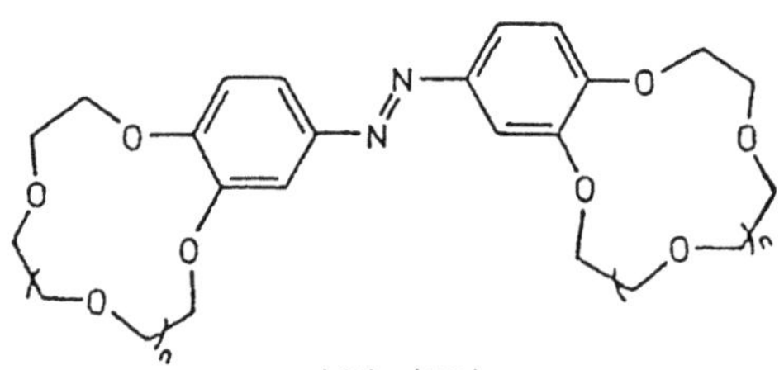

(*E*)-(**33**)

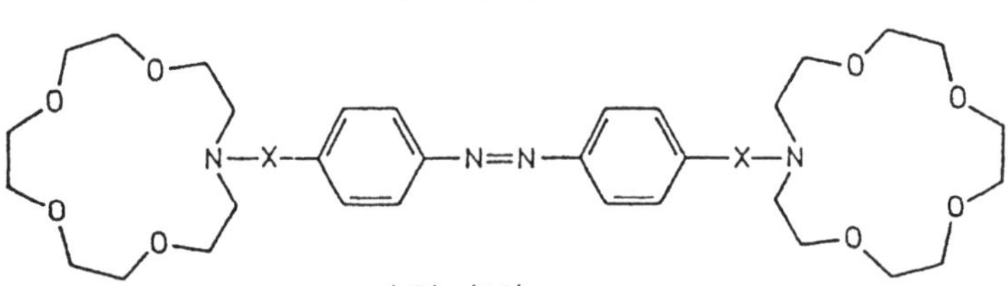

(*E*)-(**34**)

Shinkai et al. [44] evaluated the solution properties of complexes formed from (**33**: n = 3) and polymethylenediammonium cations, $H_3N^+(CH_2)_mNH_3^+$ in detail. It was found that when the distance between the two ammonium cations is shorter than that between the crown rings in (*E*)-(**33**: n = 3) (e.g. m = 6), they form a polymeric complex. When the two distances are comparable (e.g. m = 12), they form a 1:1 pseudo-cyclic complex. Photoisomerized

(*Z*)-(**33**:n = 3) showed the different aggregation mode because of the change in the distance between the two crown rings: the 1:1 complex for (*Z*)-(**33**:n = 3) + m = 6 diammonium salt and the 2:2 complex for (*Z*)-(**33**: n = 3) + m = 12 diammonium salt. This is a novel example of reversible interconversion between polymers and low molecular-weight pseudo-macrocycles.

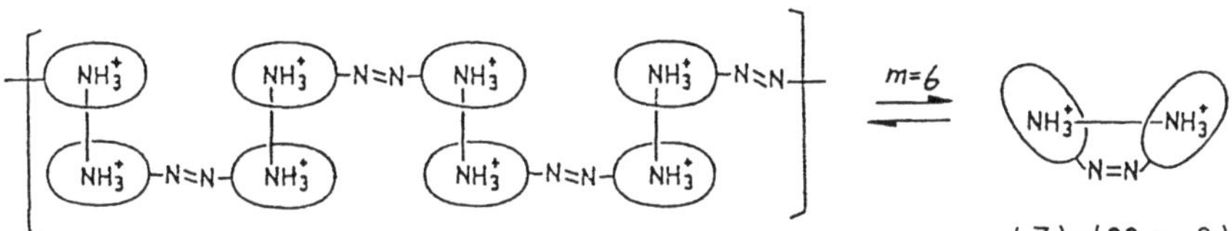

(*Z*)-(**33**; n = 3)

The change in the spacial distance between two ionophoric groups can be also achieved with the photofunctional group other than azobenzene. Irie and Kato [45] designed new photoresponsive ionophores named "molecular tweezers". As illustrated in (**35**) and (**36**) the photofunctional group is thioindigo. The (*E*)-to-(*Z*) and (*Z*)-to-(*E*) isomerization occur reversibly by irradiation of 550 nm and 450 nm light, respectively. Solvent extraction of metal cations with (**35**) revealed that the (*E*)-form has no binding ability to any of the metal cations, whereas K^+, Rb^+, and Na^+ were extracted ($Na^+ < Rb^+ < K^+$) by the photogenerated (*Z*)-form. In addition, the (*Z*)-form showed the high binding ability toward heavy metal cations such as Ag^+, Hg^+, and Cu^{2+}.

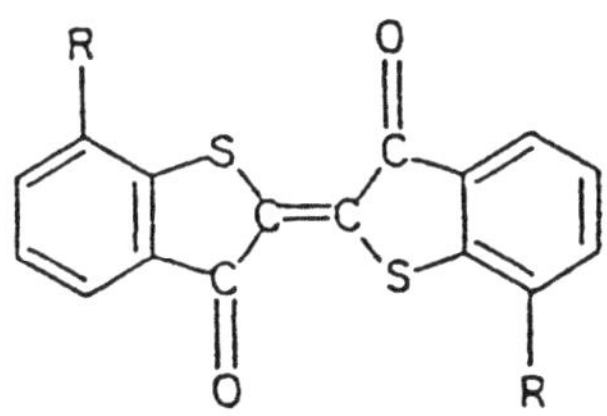

(E)-(**35**) : $R=COOCH_2CH_2OCH_3$

(E)-(**36**) : $R=CO(OCH_2CH_2)_3OCH_3$

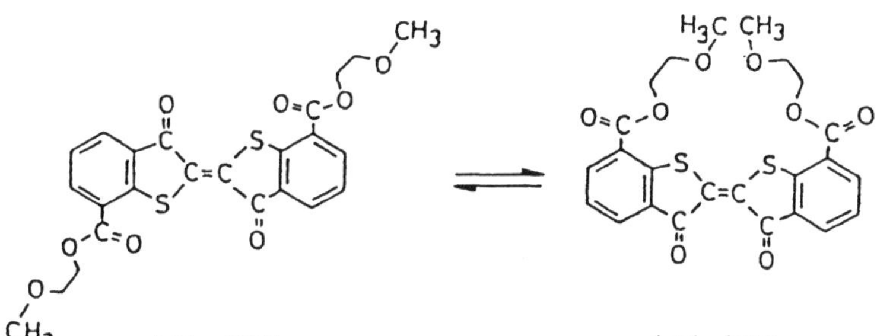

(*E*)-(**35**) (*Z*)-(**35**)

The examples for photoresponsive crown ethers described in this section suggest that in a system where two functional groups work cooperatively, the distance between these two groups is essential. In such cases, the photofunction can be controlled by the combination with the photofunctional group which changes the distance.

2.3.4 Crown Ethers Capped with Anionic or Cationic Groups

The basic concept described in the section on bis(crown ethers) can be extended to the molecular design of crown ethers with a photoresponsive ionic cap. As mentioned in the section on pH responsive crown ethers, the anionic group can act cooperatively with the crown ring upon extraction of alkali and alkaline earth metal cations. In order to exert the cooperativity, the anionic group should be placed exactly on the top of the crown ring. Thus, the possibility arises that the ion affinity is controllable by the change in the spacial position of the anionic group. For example, if one can synthesize a crown ether putting on and off an anion-cap in response to photoirradiation, it would lead to the new photocontrol of the ion binding ability.

(37) is designed so that the phenolate anion can move to the top of the crown ring upon photoisomerization of the azobenzene segment from the (*E*)-form to the (*Z*)-form [46]. The nitro group and the *n*-butyl group are introduced to lower the pK_a of the phenol group and to enhance the lipophilicity, respectively. Two-phase solvent extraction with **(37)** showed that the extractability is

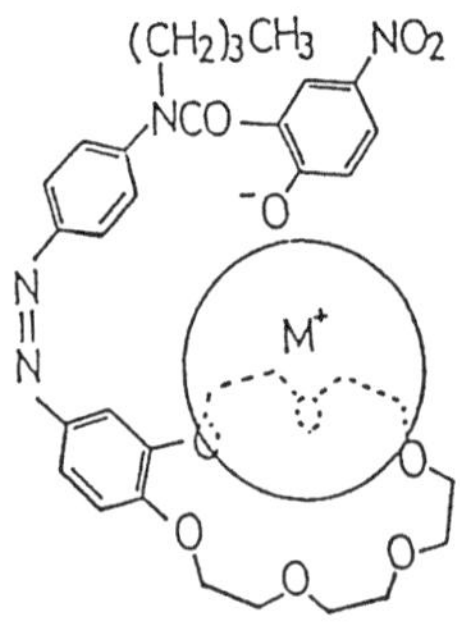

(*Z*)-(**37**)

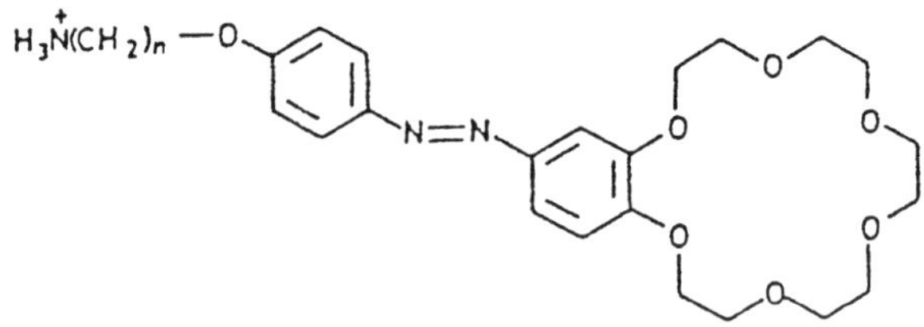

(*E*)-(**38**)

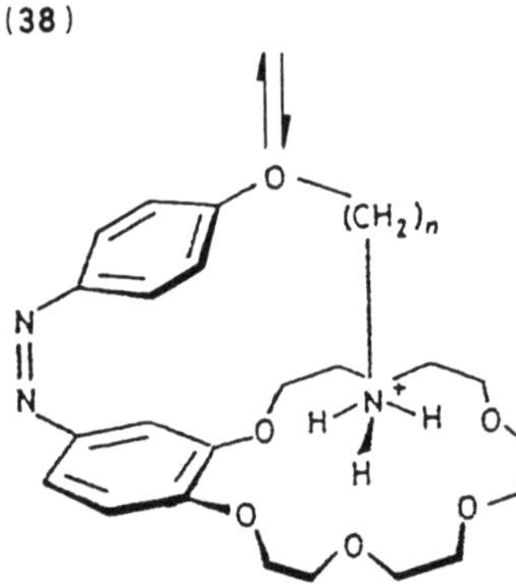

(*Z*)-(**38**)

markedly improved by UV light irradiation indicating that the crown and the cap act cooperatively in the (Z)-form. In particular, K^+, Rb^+, and Ca^{2+} were extracted efficiently. The result clearly indicates that the ion affinity is actually changed by the change in the spacial position of the phenolate anion.

(**38**:n = 4, 6, 10) have a crown ether ring and an ammonium alkyl group attached to the two sides of an azobenzene [47]. These crown have been designed so that intramolecular "biting" of the ammonium group on to the crown can only occur upon photoisomerization to the (Z)-form. Examination of the CPK model suggested that such "tail-biting" can really occur in (Z)-(**38**:n = 6, 10) but is not the case for (Z)-(**38**:n = 4) because the spacer in (Z)-(**38**:n = 4) is too short to effect the "tail-biting". This view was evidenced by the thermal (*Z*)-to-(*E*) isomerization: the first-order rate constants were much smaller than those for the free amine counterparts. This suggests that the ammonium tail in (Z)-(**38**:n = 6, 10) interacts intramolecularly with the crown ether ring. In solvent extraction the metal affinity for n = 6 and 10 was markedly reduced by UV-light irradiation and in particular the affinity with K^+ almost disappeared. This means that the intramolecular ammonium complexation occurs in preference to the intermolecular metal complexation. On the other hand, the metal affinity for n = 4 was less affected by UV light irradiation. In fact, (Z)-(**38**:n = 4) still extracted K^+.

2.3.5 Ion Transport Mediated by Photoresponsive Crown Ethers

Cations are known to be transported through membranes by synthetic macrocyclic polyethers as well as by antibiotics. When the rate-determining step is the ion extraction from the IN aqueous phase to the membrane phase, the transport rate increases with increasing stability constant. On the other hand, when the rate-determining step is the ion-release from the membrane phase to the OUT aqueous phase, the carrier must reduce the stability constant in order to attain the efficient decomplexation. As a result, the maximum rate of ion transport is observed for the ion carriers with the moderate binding constant (K_s approx. $10^6\ M^{-1}$) [48, 49]. As described in the Introduction, some polyether antibiotics feature the interconversion between the cyclic and noncyclic forms in the membrane, a feature by which the transport system escapes from the limitation of the rate-determining step. Here, an idea arises: provided that the ion-binding ability of the carrier can be changed by light, it should be possible to design light-driven ion-transport systems. This idea should be of particular importance in the system where the ion-release is rate-determining.

Shinkai et al. [38] examined the photoirradiation effect on (**32**:n = 1)-mediated Na^+ transport across a model organic membrane. Permeation of Na^+ was scarcely detected in the dark. In contrast, Na^+ was transported under UV light irradiation and the rate was faster than that mediated by dibenzo-18-crown-6. In this system Na^+ extraction from the IN aqueous phase to the liquid membrane phase is rate-limiting and the extraction speed is improved by the photochemical formation of (Z)-(**32**:n = 1).

In K^+ transport with (**33**:n = 2) across a liquid membrane, it was found that the rate is accelerated by the UV-light irradiation which mediates the (*E*)-to-(*Z*) isomerization [41]. Thus, the rate enhancement is attributed to the increased extraction speed by (*Z*)-(**33**:n = 2) from the IN aqueous phase to the membrane phase. Interestingly, the rate was further enhanced by alternate irradiation by UV and visible (which mediates the (*Z*)-to-(*E*) isomerization) light (Fig. 1) [50]. The finding is rationalized in terms of the increased release speed from the membrane phase to the OUT aqueous phase. Therefore, the rate-limiting step has been switched from the ion extraction to the ion release by alternate light irradiation.

(**37**) and (**38**) act as pH-responsive crown ethers as well as photoresponsive crown ethers. With the aid of pH gradient and light they can carry metal cations against the concentration gradient [46, 47]. For example, (**38**) acts as an ion carrier for active transport of K^+ from the basic IN aqueous phase to the acidic OUT aqueous phase (Fig. 2) [47]. As expected, the active transport of K^+ is efficiently speeded up by UV-light irradiation. Similarly, (**37**) having the photoresponsive anionic cap acts as an efficient ion carrier for Ca^{2+} and can concentrate Ca^{2+} in the acidic OUT aqueous phase against its concentration gradient [46].

The above examples demonstrate that the rate of carrier-mediated ion transport is controllable by an on–off light switch. The novel phenomenon has been attained by the combination of a photofunctional group with a crown ether which carries metal cations. One may regard, therefore, that the pH gradient used for ion-transport by some polyether antibiotics is replaced by light energy in the present light-driven transport systems.

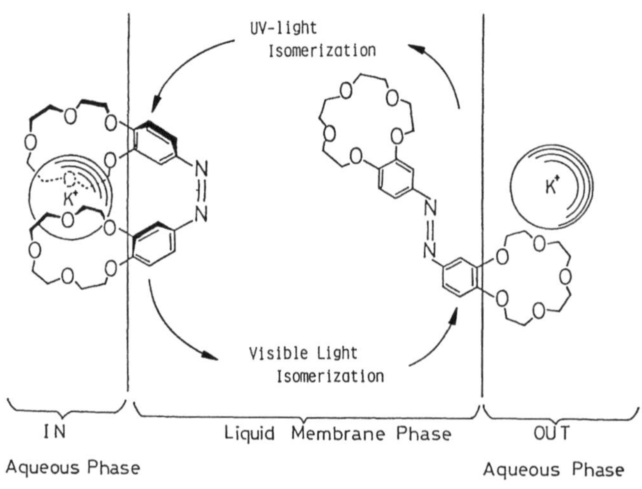

Fig. 1. Schematic representation of light-driven K^+ transport by (**33**:n = 2)

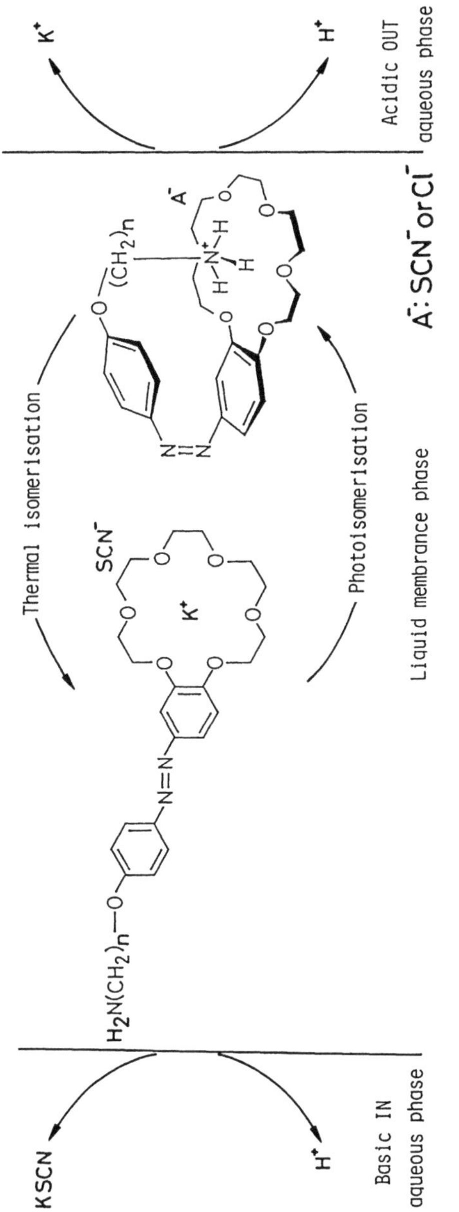

Fig. 2. Schematic representation of active K^+ transport mediated by **(38)**

3 Functionalized Calixarenes

"Calixarene" is a cavity-shaped cyclic molecule made up of a benzene unit. In contrast to the attractive architecture, the functionalized derivatives are rarely synthesized and therefore, studies of host–guest chemistry related to calixarenes have been very limited [51–53]. In fact, Gutsche stated in his review article written in 1985 that there had not been any data published in support of solution complexes of calixarenes [52]. This is in sharp contrast to cyclodextrins which can form a variety of host–guest-type solution complexes. We thus introduced appropriate functional groups into calixarenes so that they could selectively bind guest ions and molecules. We have found that calixarenes can act as a new class of catalysts, ligands, and host molecules [54, 55].

3.1 Syntheses of Fuctionalized Calixarenes

In order to introduce functional groups into calixarenes in good yields, one has to employ quantitative reactions. If the reaction results in a mixture of fully-substituted and lower-substituted materials, the purification is extremely difficult. Scheme 1 indicates a synthetic route to anionic, water-soluble calixarenes, ($\mathbf{42}_n$R) and ($\mathbf{46}_n$R) [56–58]. The key steps are ($\mathbf{40}_n$) → ($\mathbf{41}_n$) (sulfonation) and ($\mathbf{43}_n$R) → ($\mathbf{44}_n$R) (chloromethylation), which proceed almost quantitatively. In ($\mathbf{42}_n$R) and ($\mathbf{46}_n$R) anionic groups are introduced onto the upper rim (open side) of the calixarene cavity. On the other hand, treatment of ($\mathbf{39}_n$) or ($\mathbf{40}_n$) with 1,3-propane sultone gives new water-soluble calixarenes ($\mathbf{47}_n$R) which have anionic groups on the lower rim (closed side) of the calixarene cavity (Scheme 2) [59, 60]. Since the molecular architecture is quite different between ($\mathbf{42}_n$R) and ($\mathbf{47}_n$R), one can expect the different host–guest behaviors in guest inclusion.

Cationic, water-soluble calixarenes ($\mathbf{50}_n$R) and ($\mathbf{51}_n$R) are synthesized from ($\mathbf{41}_n$) via ($\mathbf{48}_n$) (Scheme 3) [61, 62] or more readily by quaternization of ($\mathbf{44}_n$R) (Scheme 4) [58]. Neutral, water-soluble calixarenes ($\mathbf{53}_n$R) are synthesized by the reaction of ($\mathbf{52}_n$R) with diethanolamine (Scheme 5) [63].

More recently, we found that the diazo-coupling reaction with calixarenes occurs readily in organic solvents in the presence of a base (e.g. pyridine: Scheme 6) [64, 65]. When X = SO_3Na or $N^+Me_3X^-$, the reaction provides new water-soluble calixarenes with an extended cavity [65]. Since the absorption spectra change in response to the association of metal cations with the OH groups, they are expected to serve as a new class of chromogenic calixarenes. Very interestingly, we unexpectedly discovered that the diazo-coupling reaction between calix[4]arene and benzenediazonium ions with electron-withdrawing *p*-substituents proceeds autoacceleratively [64]. For example, when ($\mathbf{40}_4$) and *p*-nitrobenzenediazonium tetrafluoroborate were allowed to react in THF in the presence of pyridine, unreacted ($\mathbf{40}_4$) decreased with increasing molar ratio (*p*-nitrobenzenediazonium/($\mathbf{40}_4$) = 1 ~ 4). Interestingly, tetrasubstituted ($\mathbf{54}_4NO_2$) was always recovered in high yields and mono-, di-, and tri-substi-

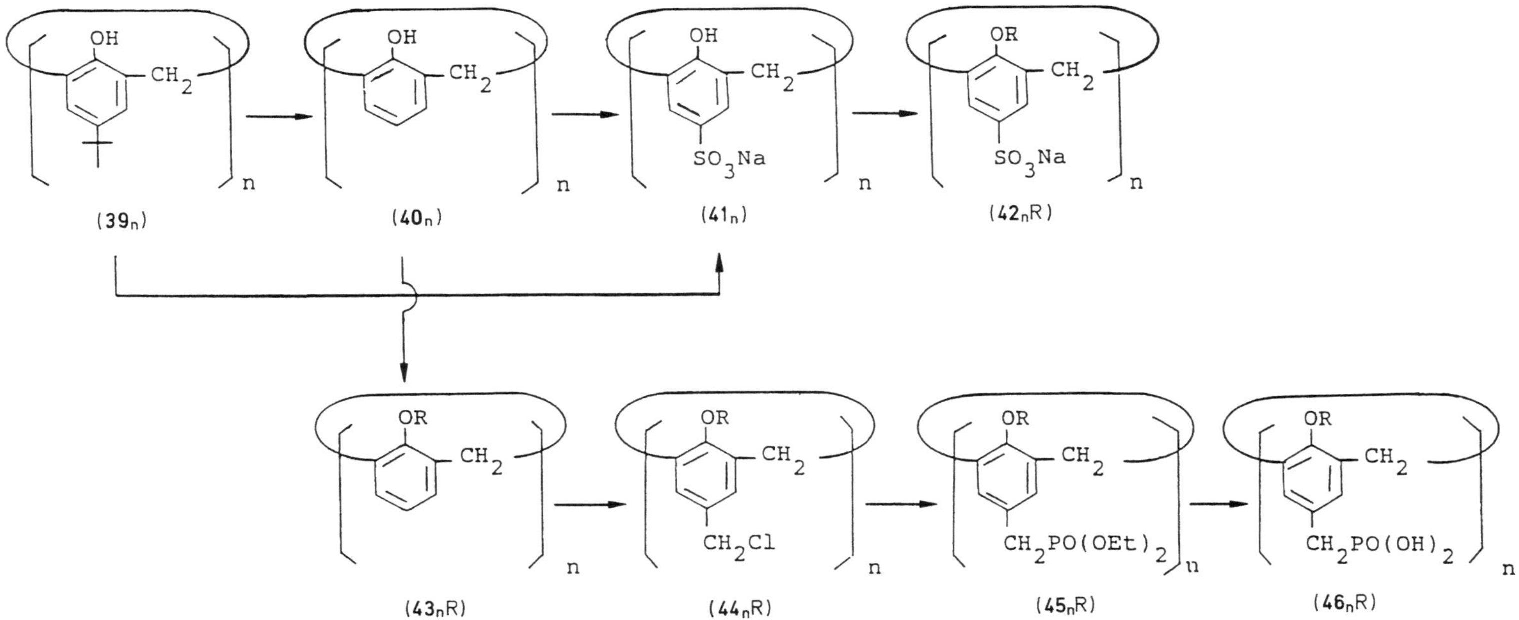
OH
CH2
(39n)
OH
CH2
(40n)
OH
CH2
SO3Na
(41n)
OR
CH2
SO3Na
(42nR)
OR
CH2
(43nR)
OR
CH2
CH2Cl
(44nR)
OR
CH2
CH2PO(OEt)2
(45nR)
OR
CH2
CH2PO(OH)2
(46nR)
n

Scheme 1

(39_n) or (40_n) 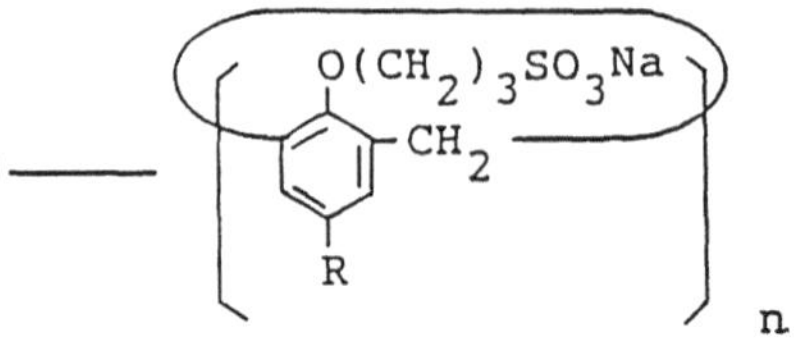

(47_nR)

Scheme 2

(41_n) 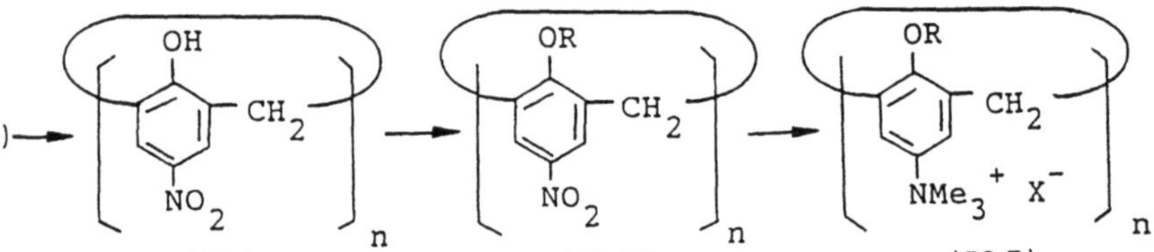

(48_n) (49_nR) (50_nR)

Scheme 3

(44_nR) 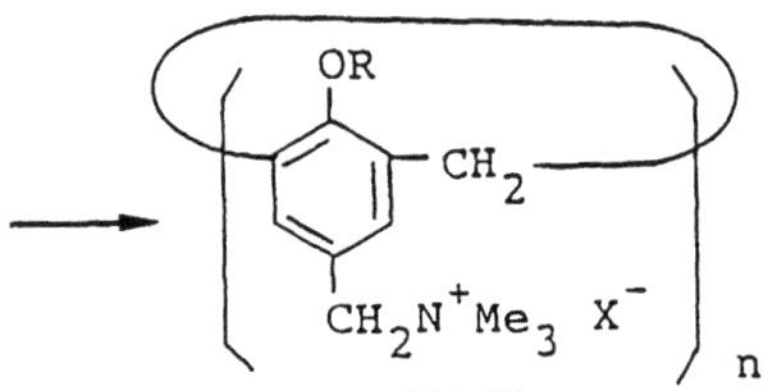

(51_nR)

Scheme 4

(42_nR) 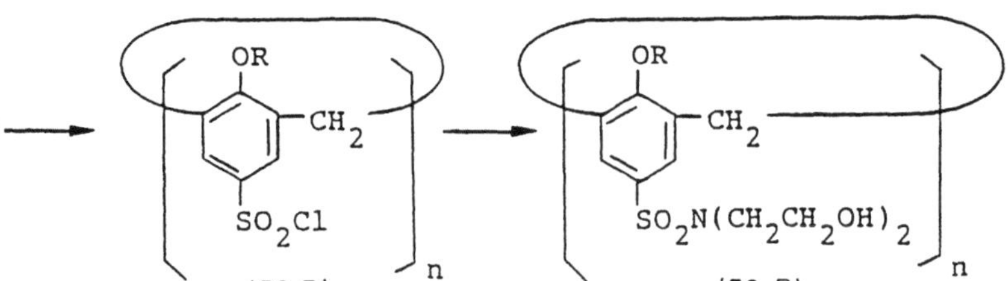

(52_nR) (53_nR)

Scheme 5

tuted products were low and almost constant. The result indicates that the diazo-coupling reaction with ($\mathbf{40}_4$) accelerates the subsequent diazo-coupling reaction. This novel autoacceleration effect was reasonably explained by the strong hydrogen-bonding among the OH groups: that is, substitution with

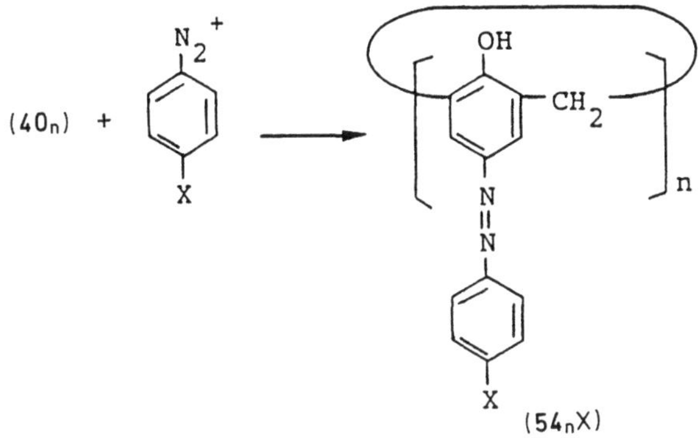

Scheme 6

electron-withdrawing diazo-groups lowers the pK_a of the neighboring phenol units and facilitates the subsequent diazo-coupling with the dissociated phenolate units.

3.2 Guest Inclusion Phenomena

Since calixarenes possess a cylindrical architecture similar to cyclodextrins, they are expected to form inclusion complexes. In fact, several groups have reported on guest inclusion in the solid state [51, 52, 66–71]. In contrast, the data related to solution complexes have been very limited [53]. We found that in an aqueous system, water-soluble calixarenes, shown in Schemes 1–5, form complexes with a variety of organic guest molecules [54–63, 72].

Phenol Blue acts as a solvent property indicator: the absorption maximum (658 nm in water) shifts to shorter wavelengths in nonpolar solvents (e.g. 552 nm in cyclohexane). Measurements of the absorption maximum in the presence of (**41**$_n$) and (**42**$_n$R) established that i) the absorption maximum in the presence of (**41**$_6$) shifts to the longer wavelength (685 nm) and ii) the plot for (**42**$_6$Dod) is biphasic, a large blue shift (by 66 nm) followed by a relatively small blue shift (by 28 nm) [56, 57]. The red shift (by 27 nm) observed for (**41**$_6$) implies that apparently, (**41**$_6$) can provide a reaction field more polar than water. Since (**42**$_6$Me) cannot induce such a large red shift, the finding supports the view that Phenol Blue (PB) is bound into the cavity of (**41**$_6$) and the OH groups stabilize the change-separated excited state ($PB^{\pm}$) through hydrogen-bonding [57].

PB $PB^{\pm}$

The guest selectivity based on the calixarene cavity was investigated below the CMC [57–60]. We employed pyrene as a guest molecule because the fluorescence intensity of the first band (382 nm) decreased sensitively with increasing calixarene concentrations [59]. The stoichiometry estimated by the molar ratio method indicated the formation of a 1:1 complex with (**42**$_n$R) and (**47**$_n$R). It was found that the K values for the (**42**$_n$R) series increase dramatically (326 fold) on going from R = Me to R = Bu^n but do not increase any more on going from R = Bu^n to R = Hex. In order to fully wrap pyrene in the calixarene cavity, the attachment of *n*-butyl groups suffices. Calixarenes (**42**$_n$Bu^n) having alkyl groups on the "lower rim" result in the K (3.5–5.6) $\times$ 10^6 M^{-1} greater than (**47**$_n$R) having alkyl groups on the "upper rim" (6.3 $\times$ 10^5 M^{-1}). Since the "lower rim" is a closed side of the calixarene cavity, incorporated alkyl groups are able to act cooperatively as a binding site (Fig. 3). In contrast, alkyl groups incorporated onto the "upper rim" (open side) are so separated from each other that the cooperative action as a binding site is rather difficult. Instead, hexameric (**47**$_6$R) shows a significant selectivity toward pyrene. According to the CPK model building the molecular size of pyrene is too large for tetrameric (**47**$_4$R), too small for octameric (**47**$_8$R), and exactly fits

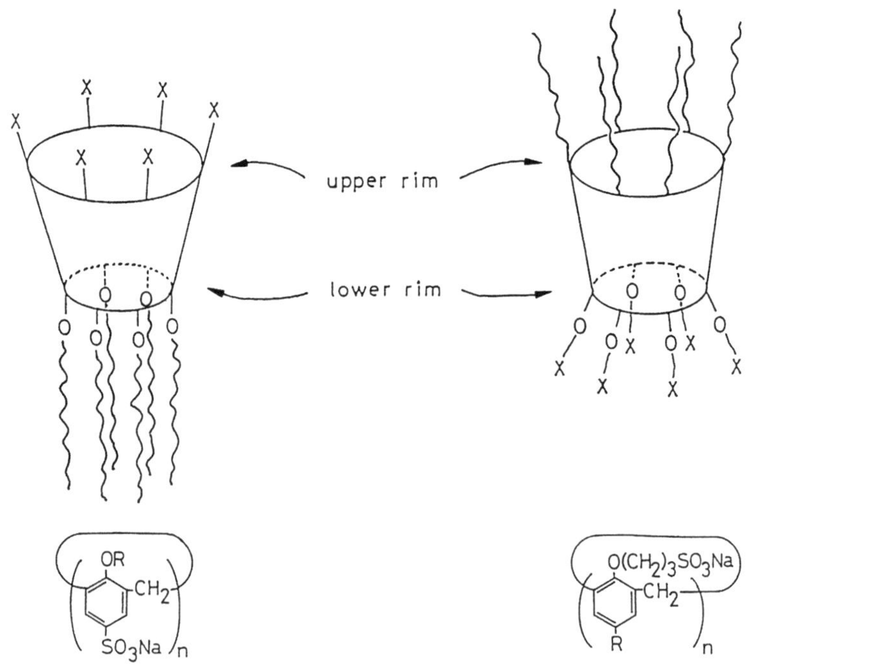

(42$_n$R) (47$_n$R)

Fig. 3. Illustration of water-soluble calixarenes having aliphatic groups on the "lower rim" (A) and the "upper rim" (B)

the cavity size of hexameric (**47**$_6$R). This implies that the selectivity rule based on the ring size is effective in (**47**$_n$R) in which *p*-substituents cannot act co-operatively [59].

Similar molecular recognition studies are reported by Gutsche and Alam [73]. They used solid–liquid extraction of aromatic guest molecules by anionic water-soluble calixarenes. Although the molecular recognition pattern on the basis of the hole-size selectivity is seen to some extent, the selectivity is not so sharp [73]. The tendency is common to all water-soluble calixarenes synthesized so far. We consider that the calixarenes, which have a more rigid framework, must be exploited in order to attain the high guest selectivity.

It is known that the rate constants for basic hydrolysis of *p*-nitrophenyl dodecanoate (*p*NPD) decrease with increasing initial ester concentration in the range of 10^{-6} to 10^{-5} M [74–76]. Evidently, *p*NPD in water forms aggregates within which the ester groups are protected from the OH^- attack and therefore hydrolyze slowly (about three orders of magnitude smaller than a short-chain monomer) [74–76]. Ammonium salts of the general structure $RNMe_3^+X^-$ are capable of disrupting the aggregates; the rate constants increase with increasing salt concentrations [76]. The rate increase is accounted for by dispersion of large aggregates into small aggregates by hydrophobic ammonium cations. A kinetic feature for this "deshielding" effect is that plots of k_{obs} (pseudo-first-order rate constants) vs [salt] curve upward [76]. We applied cationic calixarenes (**50**$_n$R) as a "deshielding" agent for this hydrolysis system. A plot of k_{obs} vs [(**50**$_4$Me)] curved upward, like conventional ammonium salts, indicating that the "deshielding" mechanism is operative [77]. In contrast, plots of k_{obs} vs [(**50**$_6$R) (R = Me, Oct)] resulted in the Michaelis-Menten-type saturation kinetics and *the rate constants were enhanced by* (1.2–5.9) × 10^5 *fold* [77]! This indicates a change in the hydrolysis mechanism from "deshielding" to "host–guest". The difference is ascribed to the cavity size and the cavity shape of calixarenes: the cavity of tetrameric (**50**$_4$R) is too small and bowl-shaped whereas that of (**50**$_6$R) is large and deep enough to accept *p*NPD.

We also found that hexameric calixarenes (**50**$_6$R) act as efficient catalysts for basic hydrolysis of phosphate esters (e.g. 2,4-dinitrophenyl phosphate), whereas tetrameric calixarenes (**50**$_4$R) scarcely catalyze the reaction [61]. Since the (**50**$_6$R)-catalyzed reaction proceeds according to the saturation kinetics, (**50**$_6$R) should form a complex with 2,4-dinitrophenyl phosphate. The association constants (K) were estimated to be 3.56 × 10^4 M^{-1} for (**50**$_6$Me) and 9.13 × 10^4 M^{-1} for (**50**$_6$Oct) at 30°C [61]. As the K values are different only by 3-fold between (**50**$_6$Me) and (**50**$_6$Oct), the complex formation would be mainly due to the electrostatic interaction between (**50**$_6$$R^{6+}$) and the ester (R'-$PO_3^{2-}$). The results suggest that three alternate cations among six placed on the upper rim of (**50**$_6$R) are complementally pre-organized for the three-point interaction with three oxygens in phosphate anions.

The K values for **41**$_n$ (n = 4, 6, and 8) were determined in D_2O by the NMR method in order to estimate the hole-size selectivity possibly operating in water-soluble calixarenes [78, 79]. Although the NMR method is more complicated than other spectroscopic methods, it is applicable to a variety of guest molecules and provides many lines of useful information. The guest molecules employed are (**55**) and (**56**). The chemical shift of (**55**) and (**56**) moved to higher magnetic field with increasing calixarene concentrations [78, 79]. This indicates that they are included in the calixarene cavity. It was found on the basis of a molar ratio method that (**41**$_4$) and (**41**$_6$) form 1:1 complexes with these guest molecules, whereas (**41**$_8$) forms 1:2 (**41**$_8$)/guest complexes. The finding supports the view that calixarenes (**41**$_n$) are capable of molecular recognition on the basis of the ring size. Examination of the thermodynamic parameters for the association process (Table 1) established that the complexation with (**41**$_4$) ($\Delta S < 0$) is mainly due to the electrostatic force, whereas that with (**41**$_6$) and (**41**$_8$) ($\Delta S > 0$) is due to the hydrophobic force [78, 79]. The K values for (**56**) are greater by a factor of about 5 than those for (**17**). When the thermodynamic parameters for (**56**) are compared with those for (**55**), we notice that the K increase in (**56**) is brought about by the increase in ΔS. This is related to the strong hydrophobic interaction of the adamantyl group with the calixarene cavity.

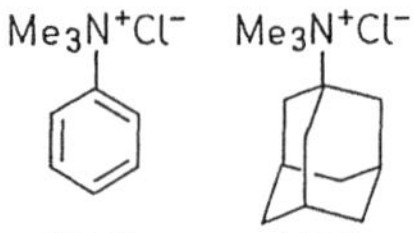

(55) (56)

What influence does included (**55**) or (**56**) have on the calixarene conformation? We estimated the influence of added (**55**) or (**56**) on T_c of (**41**$_4$) in D_2O [79, 80]. The T_c in the absence of a guest molecule appeared at 9°C. The T_c values were gradually enhanced with increasing guest concentrations and saturated at around 65°C. The CPK molecular model building and recent X-ray crystallographic studies show that the "cone" conformation provides a cavity-shaped stoma more suitable to guest-binding than the "alternate" conformation [68, 69]. It is likely, therefore, that these guest molecules act as efficient

Table 1. Association constants (K, M^{-1} at 25°C) and thermodynamic parameters (ΔH, kcal mol^{-1}; ΔS, cal mol^{-1} deg^{-1})[a]

Guest	Parameter	(**41**$_4$)	(**41**$_6$)	(**41**$_8$)	
				1:1	1:2
(**55**)	K	5600	550	5200	4600
(**55**)	ΔH	− 6.2	− 0.25	0.0	0.0
(**55**)	ΔS	− 3.6	11.7	17.0	16.7
(**56**)	K	21000	1000	19000	17000
(**56**)	ΔH	− 5.7	− 0.15	0.0	0.0
(**56**)	ΔS	0.65	13.3	19.6	19.3

[a] D_2O, 25°C, pD 7.3 with 0.1 M phosphate buffer

templates not only because of the electrostatic interaction between NMe_3^+ and oxy anions but also because of the inclusion of a phenyl or adamantyl group into the "cone"-shaped hydrophobic cavity of (**41**$_4$).

3.3 Chiral Calixarenes

In the chemistry of cyclodextrins, of particular interest is the ability of cyclodextrins to catalyze certain reactions asymmetrically, owing to the presence of the chiral cavity made up of glucose units [81, 82]. By using water-soluble calixarenes we have succeeded (in part as yet) in the molecular recognition. As a next stage, it occurred to us that the introduction of chiral substituents into calixarenes would be of great value for the development of a new class of chiral host molecules. Böhmer et al. [83] and Vicens et al. [84] reported on the syntheses of asymmetrically-substituted calix[4]arenes, but the optical isomers had never been isolated. We introduced (*S*)-2-methylbutyl groups into (**41**$_n$) and obtained chiral calix[n]arenes (**42**$_n$R*) (R* = $-CH_2CH(Me)CH_2Me$) [85, 86].

In the CD spectral measurements, (**42**$_n$R*) gave the strong Cotton effect at the ^1L_a region. The CD spectra of (**42**$_6$R*) and (**42**$_8$R*) are similar to each other but that of (**42**$_4$R*) is quite different. (**42**$_6$R*) and (**42**$_8$R*) give the first Cotton effect at 269 nm and the sign is positive. The λ_{max} values move to longer wavelengths by 23–24 nm from those of the ^1L_a band in the absorption spectra. This suggests that the Cotton effect arises from the exciton coupling of each chromophoric benzene unit. The positive sign indicates that the long molecular axes adopt the clockwise orientation when they interact at the excited state. These findings support the view that the calixarene rings of (**42**$_6$R*) and (**42**$_8$R*) are flexible enough for the exciton coupling: that is, the methylene bridges connecting two benzene chromophores can take the bond angle (ϕ) smaller than 90° (Fig. 4) [87]. In (**42**$_4$R*), on the other hand, the λ_{max} in the CD spectrum (248.5 nm) is almost equal to that of ^1L_a in the absorption spectrum (249.5 nm).

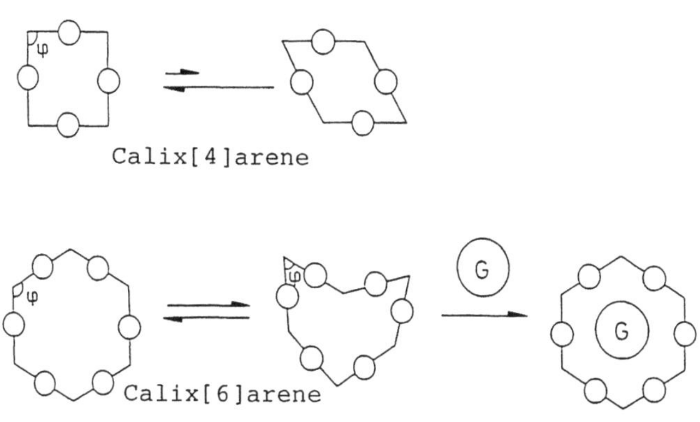

Fig. 4. Conformational fluctuation of calixarene rings

The agreement suggests the absence of the exciton coupling: that is, the methylene bridges in (**42**$_4$R*) are tense and cannot adopt the ϕ smaller than 90°.

Why is the CD spectrum of (**42**$_4$R*) so different from those for (**42**$_6$R*) and (**42**$_8$R*)? The likely answer is that introduction of alkyl substituents into the OH groups makes the calixarene ring rigid because of steric crowding. This effect appears most conspicuously in (**42**$_4$R*) with the smallest ring. In fact, the ^{1}H-NMR studies established that (**42**$_4$R*) is already fixed to "cone" because of steric inhibition of the oxygen through the annulus rotation, whereas (**42**$_6$R*) and (**42**$_8$R*) still retain the conformational freedom. This steric rigidity should suppress the conformational fluctuation in (**42**$_4$R*).

It is of great interest to study how the calixarene conformation changes upon inclusion of guest molecules. The conformational freedom of calixarenes is conveniently monitored by temperature-dependent ^{1}H NMR: for instance, the spectrum of calix[4]arene displays a singlet resonance for the $ArCH_2Ar$ methylene protons at high temperature and a pair of doublets at low temperature [51–53]. Thus, a coalescence temperature (T_c) appears at intermediary temperatures, which serves as a measure of the "cone" stability. We found that the T_c is enhanced upon inclusion of guest molecules [79, 80, 87]. We studied this through the CD spectral change in (**42**$_n$R*) [85–87]. We chose aliphatic alcohols as guest molecules because they have no absorption band at the ^1L_a region and are mostly commercially-available. First, we confirmed that in water ^{1}H-NMR chemical shifts of aliphatic alcohols move to the higher magnetic field in the presence of (**42**$_n$R*). This implies that these guest alcohols are included in the cavity of (**42**$_n$R*). The CD spectrum of (**42**$_4$R*) was not affected at all by the addition of guest alcohols. The result indicates that the conformational change in (**42**$_4$R*) is not induced by guest inclusion. This is accounted for by the rigidity of (**42**$_4$R*). In contrast, the CD Cotton bands of (**42**$_6$R*) and (**42**$_8$R*) were sensitively weakened by the addition of guest alcohols. The data lead to the following conclusions: (i) the decrease in the CD bands of (**42**$_6$R*) and (**42**$_8$R*) is observed for alcohols higher than 1-butanol, (ii) the K values increase with increasing chain length, and the K values for (**42**$_6$R*) are generally greater than those for (**42**$_8$R*). Why did the CD bands decrease upon addition of guest alcohols? As described above, the CD bands of (**42**$_6$R*) and (**42**$_8$R*) are ascribed to the exciton coupling. Thus, inclusion of guest alcohols suppresses the fluctuation of the calixarene rings and it becomes difficult for the two neighboring benzene chromophores to take ϕ smaller than 90°. It is thus clear that inclusion of guest alcohols makes the calixarene ring rigid. This conclusion obtained from the CD measurements is in line with that obtained from the ^{1}H-NMR measurements and is, therefore, general.

These results consistently demonstrate that the CD spectral method is a useful tool for the estimation of host–guest complexation properties in calixarene chemistry.

3.4 Ionophoric Calixarenes

It has been found that ester derivatives of carboxymethylcalix[n]arenes ($\mathbf{57_n}$) show the ionophoric nature and are capable of binding alkali and alkaline earth metal cations [51, 52, 88–92]. This is ascribed to interactions between "hard" oxygen bases and "hard" alkali and alkaline earth metal cations as observed for a crown ether family. In particular, tetrameric ($\mathbf{57_4}$) exhibits the markedly high Na^+ selectivity whereas hexameric ($\mathbf{57_6}$) and octameric ($\mathbf{57_8}$) exhibit the rather broad ion selectivity for K^+, Rb^+, and Cs^+. The high ion selectivity observed only for ($\mathbf{57_4}$) is probably related to the conformational rigidity. Arduini et al. [89] proposed that Na^+ is "encapsulated" in the cavity constructed with four ionophoric $-OCH_2COO-$ groups (Fig. 5). Here, two questions occurred to us which are both related to the essential behaviors of ($\mathbf{57_n}$) as ionophores: that is, if calixarenes ($\mathbf{57_n}$) really form the "encapsulated" calixarenes, then i) do they form only 1:1 metal/($\mathbf{57_n}$) complexes but not 1:2 metal/($\mathbf{57_n}$) sandwich complexes as seen for certain crown ethers? and ii) do they show characteristics of solvent-separated ion pairs? Recently, Inone et al. [93] suggested an idea that the bathochromic shift of the absorption band of the picrate anion, extracted into the organic phase with a macrocyclic ligand from aqueous metal picrate solutions, serves as a convenient measure for evaluating the ion pair tightness in solution. We thus studied the spectroscopic behaviors of alkali picrates (M^+Pic^-) in tetrahydrofuran (THF) [94].

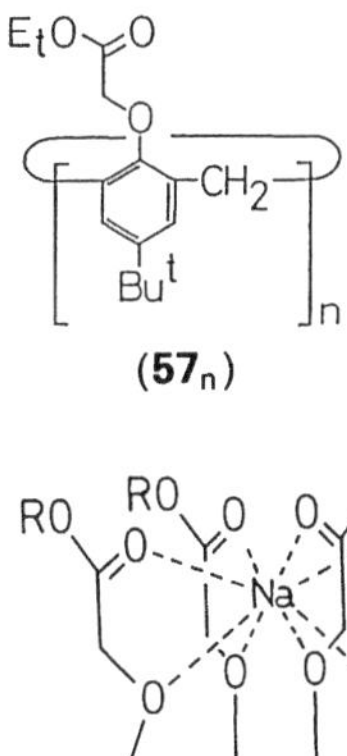

($\mathbf{57_n}$)

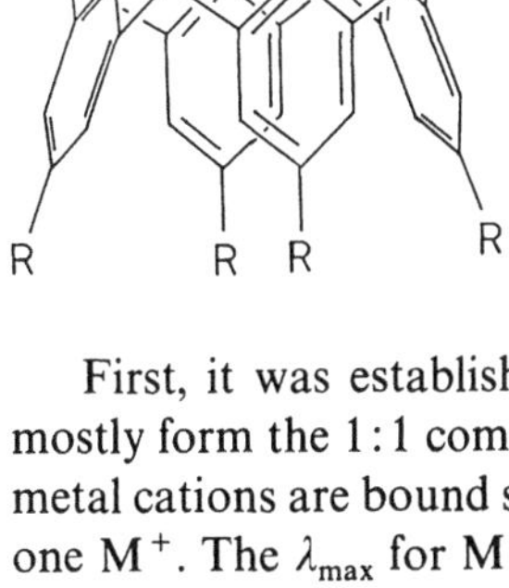

Fig. 5. Na^+ encapsulated in ($\mathbf{57_4}$)

First, it was established by a spectroscopic method that calixarenes ($\mathbf{57_n}$) mostly form the 1:1 complexes with alkali metal cations [94]. This implies that metal cations are bound so deeply in the cavity that two ($\mathbf{57_n}$)'s cannot sandwich one M^+. The λ_{max} for M^+Pic^- shifted to longer wavelengths when calixarenes

(**57_n**) were added. The largest shift (31 nm) was attained for (**57_4**)-Na^+. This shift is comparable with that induced by cryptand 222 and Na^+ (31 nm) and even greater than that induced by 18-crown-6 (29 nm for Na^+ and 13 nm for K^+). This finding supports that Na^+Pic^- bound to (**57_4**) is significantly solvent-separated. The large bathochromic shifts were also observed for K^+, (**57_6**)-Cs^+ (20–25 nm). In contrast, the shifts induced by (**57_8**) were generally small (less than 11 nm). Presumably, even though M^+ is entrapped in the cavity of (**57_8**), the M^+–Pic^- interaction can still exist because of the large flexible cavity.

In order to obtain a further insight into the (**57_n**)$\cdot M^+Pic^-$ ion pairs, we measured the ^{23}Na-NMR in THF/THF-d_8 mixed solvent at 30°C [94]. We unexpectedly found that the peak for Na^+ (added as M^+Pic^- (6.8×10^{-3} M)) is invariably broadened in the presence of (**57_n**): for example, the T_2 values are 7.08×10^{-4} s in the absence of (**57_n**), 1.62×10^{-5} s in (**57_4**), 2.11×10^{-5} in (**57_6**), and 3.18×10^{-5} s in (**57_8**). Such a broadening effect was not found to 18-crown-6 ($T_2 = 8.85 \times 10^{-4}$ s) and cryptand 222 ($T_2 = 1.47 \times 10^{-3}$ s) [94]. Supposedly, the broadening effect is related to the "encapsulation" effect that causes the slow association–dissociation rate for the (**57_n**)$\cdot Na^+$ complex.

In 1973, Gokel and Cram [95] found that crown ethers of appropriate dimensions can solubilize several arenediazonium salts in nonpolar solvents (e.g. chloroform). Subsequent spectroscopic studies established that solubilization is achieved by complexation, the linear Ar–$N^+ \equiv N$ inserting into the cavity of the crown ring [95]. The thermal decomposition of arenediazonium salts is slower when the salts are complexed by crown ethers [96]. Two opposing stabilization mechanisms have been proposed: that is, the thermal decomposition is suppressed because of (i) reduction of the positive charge at the diazo group through interaction with oxygen lone pairs or (ii) the macrocyclic effect, which sterically inhibits dediazoniation proceeding from the linear Ar–$N^+ \equiv N$ inserted in the crown ring to the bulky π-intermediate (**58**) [97]. To investigate whether arenediazonium salts are also complexed and stabilized by (**57_n**), we used (**59**) which serves as a spectroscopic probe: compound (**59**) has push–pull-type substituents, so that the absorption maximum should reflect sensitively the reduction of the positive charge at the diazo group [98]. The absorption maximum of (**59**) (680 nm in tetrachloroethane) shifted to shorter wavelengths with increasing (**57_n**) concentrations. A similar trend was also observed for crown ethers. This blue shift implies that intramolecular charge-transfer from Me_2N to N_2^+ is suppressed because of the reduction of the positive charge at the diazo group. Among them, the large blue shifts were particularly observed for (**57_6**) (123 nm) and 18-crown-6 (115 nm) [98]. This suggests that the K^+-selective ionophores can associate strongly with the diazo group. Interestingly, the dediazoniation rate of (**60**), which was efficiently suppressed by 18-crown-6, was scarcely affected by the addition of (**57_6**): that is, even though (**57_6**) can strongly complex arenediazonium salts, it cannot inhibit dediazoniation. This trend supports the view that the mechanism (ii) is responsible for the inhibition effect of 18-crown-6. (**57_6**) having an ionophoric but flexible ring cannot exert the

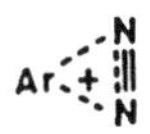

(58)

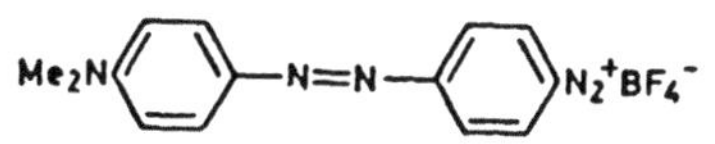

(59)

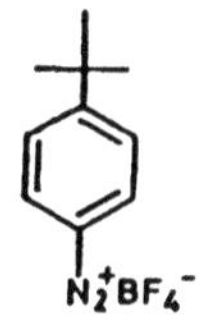

(60)

steric inhibition effect on the process from linear Ar–$N^+ \equiv N$ to the bulky π-intermediate (**58**).

Molecular recognition and chemistry at the interface are two major fields in chemistry that are bound to expand rapidly in the coming years. Since calixarenes (**57_n**) exhibit the metal recognition ability, it would be interesting to test if the ability is reproduced at the air–water interface. Regen and co-workers [99] recently reported on perforated monolayers from calix[6]arenes but the molecular recognition by calixarenes at the air–water interface has never been reported. We found that calixarenes (**57_n**) give stable monolayers: the pressure-area (π-A) isotherms of (**57_n**) are those characteristic of the monolayer formation [100]. The molecular areas were estimated to be 1.16 nm^2 for (**57_4**) and 2.06 nm^2 for (**57_6**). These values are in accord with the areas of the large sides (the upper rim) of the "cone"-shaped calixarene cavity. It is likely, therefore, that calixarenes (**57_n**) adsorbed at the air–water interface adopt the "cone" conformation, extending hydrophilic ester groups into water and hydrophobic *p–t*-butylphenyl groups into air. Calix[8]arene is a flexible molecule which can assume various conformations. Apparently, (**57_8**) changes its conformation with increasing pressures, resulting in the broad π-A behavior. Significantly, these monolayers can "respond" to alkali metal cations added to the aqueous subphase. The monolayer of (**57_4**) is much expanded on aqueous NaCl (1.0 M), but not on aqueous KCl (1.0 M) and LiCl (1.0 M). On the contrary, the monolayer of (**57_6**) is expanded when KCl is added to the subphase. The monolayer of (**57_8**) is affected by NaCl and KCl only weakly. The ion selectivity determined by two-phase solvent extraction is $Li^+ < Na^+ > K^+ > Rb^+$ for (**57_4**), $K^+ > Rb^+ > Na^+ > Li^+$ for (**57_6**), and $Rb^+ > K^+ > Na^+ > Li^+$ for (**57_8**) [89–91]. Although (**57_4**) shows the sharp selectivity toward Na^+, that of (**57_6**) and (**57_8**) is not so outstanding [89–91]. In contrast, not only (**57_4**) but also (**57_6**) exhibits a very sharp, all-or-nothing metal selectivity at the air–water interface. Clearly, the metal recognition ability of (**57_n**) is much improved when they are assembled as a monolayer. As described above, if calixarenes (**57_n**) extend hydrophilic ester groups into water and hydrophobic *p–t*-butylphenyl groups into air at the air–water interface, the conformation should be fixed to "cone". Although this view is still a matter of discussion, we believe that the fixation of the calixarene conformation is responsible for the improved metal selectivity. This suggests that by freezing the calixarene conformation the high metal selectivity would be effected even in solvent extraction.

We wish to apply ionophoric calixarenes for the selective binding of more precious metal ions. The selective extraction of uranium from sea water has attracted extensive attention from chemists because of its importance in relation to energy problems. In order to design such a ligand that can selectively extract uranyl ion (UO_2^{2+}), one has to overcome a difficult problem: i.e., the ligand must strictly discriminate UO_2^{2+} from other metal ions present in great excess in sea water. For example, the concentration of UO_2^{2+} in sea water is 3 ppb while those of competing metal cations are about 10 ppm order. Hence, the practical uranophile is required to have the UO_2^{2+} selectivity, at least, greater than 10^4. A possibly unique solution to this difficult problem is provided by the unusual coordination structure of UO_2^{2+} complexes. X-ray crystallographic studies have established that UO_2^{2+} complexes adopt either a pseudoplanar pentacoordinate or hexacoordinate structure, which is quite different from the coordination structures of other metal ions. This suggests that a macrocyclic host molecule having a nearly coplanar arrangement of either five or six ligand groups would serve as a specific ligand for UO_2^{2+} (i.e., as a practical uranophile). This approach has been investigated by several groups [101–103]. For example, Tabushi et al. [102] synthesized a macrocyclic host molecule (**61**) having six carboxylate groups in a ring. Although the stability constant for (**61**) and UO_2^{2+} is pretty high ($\log K_{uranyl} = 16.4$ at pH 10.4 and 25°C), the selectivity for UO_2^{2+} is not satisfactory (e.g. $K_{uranyl}/K_{M^{n+}} = 80$–210 for Ni^{2+} and Zn^{2+}) and the synthesis is not easy [102].

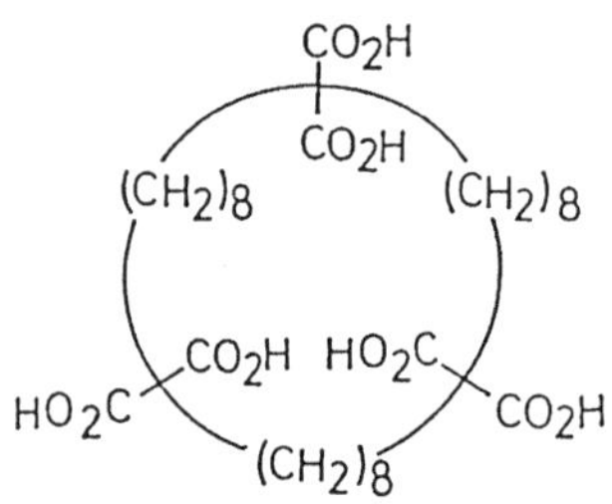

(**61**)

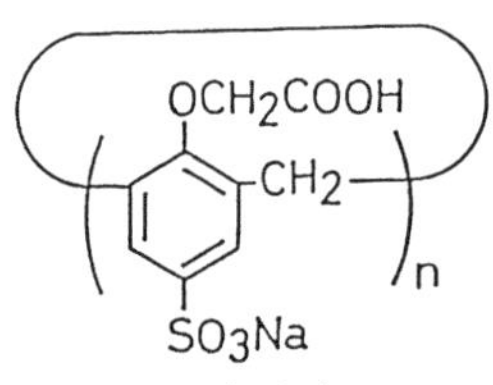

(**62**$_n$)

In the course of our studies on calixarenes, we noticed that calix[5]arene and calix[6]arene may have ideal architectures for the design of uranophiles, because introduction of ligand groups into each benzene unit of these calixarenes exactly provides the required pseudoplanar penta- and hexacoordinate structures [104, 105]. We thus applied (**41**$_n$) and (**62**$_n$) as uranophiles. We found that as shown in Tables 2 and 3, (**41**$_5$), (**41**$_6$), (**62**$_5$), and (**62**$_6$) have not only the high, record-breaking stability constants ($\log K_{uranyl} = 18.4$–19.2) but also an unusually high selectivity for UO_2^{2+} ($K_{uranyl}/K_{M^{n+}} = 10^{12}$–$10^{17}$) [105]. In contrast, the K_{uranyl} values for (**41**$_4$) and (**62**$_4$) were dramatically decreased: they are smaller by about 16 log units than those for the pentamers and the hexamers [105]. The high affinity is rationalized in terms of the "coordination-geometry selectivity": that is, the pentamers and the hexamers can provide the ligand groups arranged in a suitable way required for pseudoplanar penta- or hexacoordination on the edge of the calixarenes but the tetramers cannot. Similarly,

Table 2. Stability constants (K_{uranyl}) for calixarene derivatives and UO_2^{2+} (25°C)

Calixarene	pH	log K_{uranyl}
(**41**$_4$)	6.5	3.2
(**62**$_4$)	6.5	3.1
(**41**$_5$)	10.4	18.9
(**62**$_5$)	10.4	18.4
(**41**$_6$)	10.4	19.2
(**62**$_6$)	10.4	18.7
(**61**)	10.4	16.4

Table 3. Selectivity factors for UO_2^{2+} ($K_{uranyl}/K_{M}n^+$)

Calixarene	Metal (M^{n+})	log $K_M n+$	$K_{uranyl}/K_M n+$
(**41**$_6$)	UO_2^{2+}	(19.2)	1.0
(**41**$_6$)	Mg^{2+}	[a]	$>10^{17}$
(**41**$_6$)	Ni^{2+}	2.2	$10^{17.0}$
(**41**$_6$)	Zn^{2+}	5.5	$10^{13.7}$
(**41**$_6$)	Cu^{2+}	8.6	$10^{10.6}$
(**62**$_6$)	UO_2^{2+}	(18.7)	1.0
(**62**$_6$)	Mg^{2+}	[a]	$>10^{17}$
(**62**$_6$)	Ni^{2+}	3.2	$10^{15.3}$
(**62**$_6$)	Zn^{2+}	5.6	$10^{13.1}$
(**62**$_6$)	Cu^{2+}	6.7	$10^{12.0}$

[a] The $K_M n+$ is too small to be determined by the polarographic method

the high selectivity is rationalized in terms of the moderate rigidity of the calixarene skeleton: that is, (**41**$_6$) and (**62**$_6$) firmly maintain the pseudoplanar hexacoordination geometry and cannot accommodate either to the tetrahedral or to the octahedral coordination geometry. We also found that (**46**$_6$Me) with six phosphonate groups acts as an excellent uranophile comparable with (**41**$_6$) and (**62**$_6$) [106].

Based on the foregoing results, we applied hexacarboxylate derivatives of *p*-*t*-butylcalix[6]arene (**63**$_6$Bu^t) and *p*-*n*-hexylcalix[6]arene (**63**$_6$Hex) to solvent extraction of UO_2^{2+} from water to organic solvent (*o*-dichlorobenzene) [107]. Although the high UO_2^{2+} affinity was reproduced in solvent extraction, the UO_2^{2+} selectivity attained with (**41**$_6$) and (**62**$_6$) was not observed: that is,

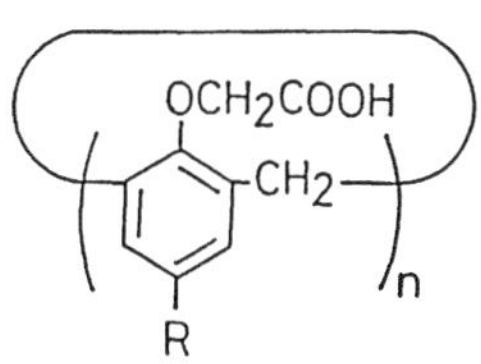

(**63**$_n$R)

competing metal cations such as Ni^{2+} and Zn^{2+} were extracted into the organic phase with UO_2^{2+}. The detailed examination of extracted species established that these competing metal cations are adsorbed to the anionic $(\mathbf{62}_6R)^{n-} \cdot UO_2^{2+}$ complex (where n = 4–6) as countercations [107]. In fact, the UO_2^{2+} selectivity was sufficiently improved by the addition of tri-*n*-octylmethylammonium cation into the organic phase because metal cations can be easily displaced by this lipophilic ammonium cation. The finding suggests that calix[n]arenes (n = 5, 6) with 2- charge, which ultimately form a neutral UO_2^{2+} complex, exhibit the high UO_2^{2+} selectivity even in solvent extraction.

4 Conclusions

In the history of host–guest chemistry over the last two decades, the chemistry of cyclodextrins and crown ethers has been a focus of central interest. Why have these two macrocycles attracted extensive attention for such a long time? We have summarized their attractive points and compared them with calixarenes in Table 4 [65]. ○ denotes that the macrocycle already satisfies the requirement. △ denotes that the requirement can be easily satisfied by simple modification of the macrocycle. × denotes that it is considerably difficult to satisfy the requirement. The most important requirement for attaining molecular recognition would be if the macrocycles with different cavity size could be synthesized systematically (Entry 1). For example, it is known that calixarene-like cyclic oligomers can be also made from catechol and resorcinol, but catechol gives only the trimer and resorcinol only the tetramer. To the best of our knowledge, this requirement can be satisfied only by cyclodextrins, crown ethers, and calixarenes. Furthermore, it is possible to synthesize these macrocycles on a large scale (Entry 2). At the early stage of cyclodextrin chemistry, the research activity was seriously limited by the low yield in cyclodextrin synthesis. When cyclodextrins became easily accessible owing to the discovery of the enzymatic synthesis, the spectrum of cyclodextrin chemistry expanded explosively. This indicates the importance of Entry 2. Other entries are also important in the design of finely functionalized host molecules, which are more or less satisfied by

Table 4. Comparison of three representative host molecules

Entry	Cyclodextrin	Crown ether	Calixarene
1 Systematic change in the ring size	○	○	○
2 Large-scale preparation	○	○	○
3 Spectroscopic transparency	○	○	×
4 Neutrality under the working conditions	○	○	△
5 Optical-activity	○	△	△
6 Synthetic easiness for derivatives	△	○	○
7 Functions as ionophores	×	○	○
8 Functions as cavity-shaped hosts	○	×	○

calixarenes. We believe that "calixarenes" will soon appear on the stage of host–guest chemistry as the "third supramolecule". As a new combination of switch-functions with calixarenes or calixarene-based host molecules, we are now designing switch-functionalized calixarenes. As demonstrated in this chapter, the "switched-on" function is quite important in the system where the "guest-binding" is included in the rate-limiting step. We believe that this concept would be successfully applied to the decomplexation of encapsulated Na^+ or superuranophile · UO_2 complexes.

5 References

1. Raymond KN, Carrano CJ (1979) Acc. Chem. Res. 12: 183
2. Choy EM, Evans DF, Cussler EL (1974) J. Am. Chem. Soc. 96: 7058
3. Shinkai S, Manabe O (1974) Top. Curr. Chem. 121: 67
4. Yamazaki N, Hirano A, Nakahama S (1979) J. Macromol. Sci.-Chem. A13: 321
5. Gardner JO, Beard CC (1978) J. Med. Chem. 21: 357
6. Taguchi K, Hiratani K, Sugihara H, Ito K: Chem. Lett. 1984: 1457
7. Matsushima K, Kobayashi H, Nakatsuji Y, Okahara M: Chem. Lett. 1983: 701
8. Nakatsuji Y, Kobayashi H, Okahara M (1986) J. Org. Chem., 51: 3789
9. Nakatsuji Y, Kobayashi H, Okahara M: J. Chem. Soc. Chem. Commun. 1983: 800
10. Shinkai, S, Kinda H, Araragi Y, Manabe O (1983) Bull. Chem. Soc. Jpn 56: 559
11. Wierenga W, Evans BR, Woltersom JA (1979) J. Am. Chem. Soc. 101: 1334
12. Strzelbicki J, Bartsch RA (1982) J. Membr. Sci. 10: 35
13. Charewicz WA, Bartsch RA (1983) J. Membr. Sci. 12: 323
14. Fyles TM (1986) J. Chem. Soc., Faraday Trans. I 82: 617
15. Sugihara K, Kamiya H, Yamaguchi M, Kaneda T, Misumi S (1981) Tetrahedron Lett. 22: 1619
16. Kaneda T, Sugihara K, Kamiya H, Misumi S (1981) Tetrahedron Lett. 22: 4407
17. Wolf RE Jr, Cooper SR (1984) J. Am. Chem. Soc. 106: 4646
18. Bock H, Hierholzer B, Vögtle F, Hollmann G (1984) Angew. Chem. Int. Ed. Engl. 23: 57
19. Hayakawa K, Kido K, Kamematsu K: J. Chem. Soc., Chem. Commun. 1986: 269
20. Maruyama K, Sohmiya H, Tsukube H (1985) Tetrahedron Lett., 26: 3583
21. S. Nakatsuji, Ohmori Y, Iyoda M, Nakashima K, Akiyama S (1983) Bull. Chem. Soc. Jpn. 56: 3185
22. Shinkai S, Inuzuka K, Miyazaki O, Manabe O (1984) J. Org. Chem. 49: 3440
23. Shinkai S, Inuzuka K, Miyazaki O, Manabe O (1985) J. Am. Chem. Soc. 107: 3950
24. Raban M, Greenblatt J, Kandil F: J. Chem. Soc., Chem. Commun. 1983:1409
25. Shinkai, Inuzaka K, Hara K, Sone T, Manabe O (1984) Bull. Chem. Soc. Jpn. 57: 2150
26. Shinkai S, Minami T, Araragi Y, Manabe O: J. Chem. Soc., Perkin Trans. I 1985: 503
27. Shinkai S, Ogawa T, Nakaji T, Kusano Y, Manabe O: Tetrahedron Lett. 1979: 4569
28. Shinkai S, Nakaji T, Nishida Y, Ogawa T, Manabe O (1980) J. Am. Chem. Soc. 102: 5860
29. Ammon HL, Bhattecharjee SK, Shinkai S, Honda Y (1984) J. Am. Chem. Soc. 106: 262
30. Shinkai S, Kouno T, Kusano Y, Manabe O: J. Chem. Soc., Perkin Trans. I 1982: 2741
31. Shinkai S, Honda Y, Ueda K, Manabe O (1984) Bull, Chem. Soc. Jpn. 57: 2144
32. Shinkai S, Miyazaki K, Nakashima M, Manabe O (1985) Bull. Chem. Soc. Jpn. 58: 1059
33. Desvergne J-P, Bouas-Laurent H: J. Chem. Soc., Chem. Commun. 1978: 403
34. Bouas-Laurent H, Castellan A, Desvergne J-P (1980) Pure Appl. Chem. 52: 2633
35. Bouas-Laurent H, Castellan A, Daney M, Desvergne J-P, Guinand G, Marsan P, Riffand M-H (1986) J. Am. Chem. Soc., 108: 315
36. Yamashita I, Fujii M, Kaneda T, Misumi S, Otubo T (1980) Tetrahedron Lett. 21: 541
37. Shinkai S, Minami T, Kusano Y, Manabe O (1983) J. Am. Chem. Soc. 105: 1851
38. Shinkai S, Miyazaki K, Manabe O: J. Chem. Soc. Perkin Trans. I 1987: 449.
39. Kimura K, Tamura H, Tsuchida T, Shono T: Chem. Lett. 1979: 611
40. Shinkai S, Ogawa T, Kusano Y, Manabe O: Chem. Lett. 1980: 283
41. Shinkai S, Nakaji T, Ogawa T, Shigemetsu K, Manabe O (1981) J. Am. Chem. Soc. 103: 111
42. Shinkai S, Shigematsu K, Kusano Y, Manabe O: J. Chem. Soc., Perkin Trans. I 1981: 3279

43. Shinkai S, Ogawa T, Kusano Y, Manabe O, Kikukawa K, Goto T, Matsuda T (1982) J. Am. Chem. Soc. 104: 1960
44. Shinkai S, Yoshida T, Manabe O, Fuchita Y: J. Chem. Soc. Perkin Trans. I 1988: 1431
45. Irie M, Kato M (1985) J. Am. Chem. Soc. 107: 1024
46. Shinkai S, Minami T, Kusano Y, Manabe O (1982) J. Am. Chem. Soc. 104: 1967
47. Shinkai S, Ishihara M, Ueda K, Manabe O: J. Chem. Soc., Perkin Trans. II 1985: 511
48. Kirch M, Lehn J-M (1975) Angew. Chem. Int. Ed. Engl. 14: 555
49. Lamb JD, Christensen JJ, Oscarson JL, Asay BW, Izatt RM (1980) J. Am. Chem. Soc. 102: 6820
50. Shinkai S, Shigematsu K, Sato M, Manabe O: J. Chem. Soc., Perkin Trans. I 1982: 2735
51. Gutsche CD (1983) Acc. Chem. Res. 16: 161
52. Gutsche CD (1984) Top. Curr. Chem. 123: 1; Gutsche CD (1985) Host guest complex chemistry/Macrocycles, Springer, Berlin Heidelberg New York, p 375
53. Bauer LJ, Gutsche CD (1985) J. Am. Chem. Soc. 107: 6063
54. Shinkai S (1986) Pure Appl. Chem. 58: 1523
55. Shinkai S, Manabe O: Nippon Kagaku Kaishi 1988: 1917
56. Shinkai S, Mori S, Tsubaki T, Sone T, Manabe O (1984) Tetrahedron Lett. 25: 5315
57. Shinkai S, Mori S, Koreishi H, Tsubaki T, Manabe O (1986) J. Am. Chem. Soc. 108: 2409
58. Arimura T. Nagasaki T, Shinkai S, Matsuda T (1989) J. Org. Chem. 54: 3766
59. Shinkai S, Kawabata H, Arimura T, Matsuda T, Satoh H, Manabe O: J. Chem. Soc., Perkin Trans. I 1989: 1073
60. Shinkai S, Arimura T, Araki K, Kawabata H, Satoh H, Tsubaki T, Manabe O, Sunamoto J: J. Chem. Soc., Perkin Trans. I 1989: 2039
61. Shinkai S, Shirahama Y, Tsubaki T, Manabe O: J. Chem. Soc., Perkin Trans. I 1989: 1859
62. Shinkai S, Araki K, Tsubaki T, Arimura T, Manabe O: J. Chem. Soc., Perkin Trans. I 1987: 2297
63. Kawaguchi H, Shinkai S, Manabe O 1989: 1859
64. Shinkai S, Araki K, Shibata J, Manabe O: J. Chem. Soc., Perkin Trans. I 1989: 195
65. Shinkai S, Araki K, Shibata J, Tsugawa D, Manabe O: Chem. Lett. 1989: 931
66. Andreetti GD, Ungaro R, Pochini A: J. Chem. Soc., Chem. Commun. 1979: 1005
67. Coruzzi M, Andreetti GD, Bocchi V, Pochini A, Ungaro R: J. Chem. Soc., Perkin Trans. II 1982: 1133
68. Ungaro R, Pochini A, Andreetti GD, Domiano P: J. Chem. Soc., Perkin Trans. II 1985: 197
69. Bott SG, Coleman AW, Atwood JL (1986) J. Am. Chem. Soc. 108: 1709; (1988) 110: 610
70. McKervey MA, Seward EM, Ferguson G, Ruhl BL (1986) J. Org. Chem. 51: 3581
71. Coleman AW, Bott S, Atwood JL (1986) J. Inclusion Phenom. 4: 247
72. A water-soluble calixarene (*p–t*-butylcalix[4]arene tetracarboxylic acid) was also reported by the Italian group, but the water solubility is low, especially, in the presence of salts: Arduini A, Reverberi S, Ungaro R: J. Chem. Soc., Chem. Commun. 1984: 981. More recently, Poh et al. synthesized a new water-soluble macrocycle by condensation of formaldehyde and 1,8-dihydroxynaphthalene-3,5-disulfonate: Poh BL, Lim CS, Kho KS (1989) Tetrahedron Lett. 30: 1005
73. Gutsche CD, Alam I (1988) Tetrahedron 44: 4689
74. Menger FM, Portnoy CE (1968) J. Am. Chem. Soc. 90: 1875
75. Murakami Y, Aoyama Y, Kida M: J. Chem. Soc., Perkin Trans. II 1977: 1947
76. Menger FM, Venkataram UV (1986) J. Am. Chem. Soc. 108: 2980
77. Shinkai S, Shirahama Y, Tsubaki T, Manabe O (1989) J. Am. Chem. Soc. 111: 5477
78. Shinkai S, Araki K, Manabe O (1988) J. Am. Chem. Soc. 110: 7214
79. Shinkai S, Araki K, Matsuda T, Manabe O (1989) Bull. Chem. Soc. Jpn. 62: 3856
80. Arimura T, Kubota M, Araki K, Shinkai S, Matsuda T (1989) Tetrahedron Lett. 30: 2563
81. Breslow R (1980) Acc. Chem. Res. 13: 170
82. Tabushi I (1982) Acc. Chem. Res. 15: 66
83. Böhmer V, Marschollek F, Zetta L (1987) J. Org. Chem. 52: 3200
84. Casabianca H, Royer J, Satrallah A, Taty C, Vicens J (1987) Tetrahedron Lett. 28: 6595
85. Shinkai S, Arimura T, Satoh H, Manabe O: J. Chem. Soc., Chem. Commun. 1987: 1495
86. Arimura T, Edamitsu S, Shinkai S, Manabe O, Muramatsu T, Tashiro M: Chem. Lett. 1987: 2269
87. Shinkai S, Arimura T, Kawabata H, Hirata Y, Fujio K, Manabe O, Muramatsu T (to be submitted)
88. Ungaro R, Pochini A, Andreetti GD (1984) J. Inclusion Phenom. 2: 199

89. Arduini A, Pochini A, Reverberi S, Ungaro R, Andreetti GD, Ugozzoli F (1986) Tetrahedron 42: 2089
90. Change S-K, Cho I: Chem. Lett. 1984: 474; Chang S-K, Cho I: J. Chem. Soc., Perkin Trans. I 1986: 211
91. McKervey MA, Seward EM, Ferguson G, Ruhl B, Harris S: J. Chem. Soc., Chem. Commun. 1985: 388
92. Kimura K, Matsuo M, Shono T: Chem. Lett. 1988: 615
93. Inoue Y, Fujiwara C, Wada K, Tai A, Hakushi T: J. Chem. Soc., Chem. Commun. 1987: 393
94. Arimura T, Kubota M, Matsuda T, Manabe O, Shinkai S (1989) Bull. Chem. Soc. Jpn. 62: 1674
95. Gokel GW, Cram DJ: J. Chem. Soc., Chem. Commun. 1973: 482
96. Bartsch RA, Haddock HF, Juri PN (1976) J. Am. Chem. Soc. 98: 6753
97. Nakazumi H, Szele I, Yoshida K, Zollinger H (1983) Helv. Chim. Acta 66: 1721
98. Shinkai S, Edamitsu S, Arimura T, Manabe O: J. Chem. Soc., Chem. Commun. 1988: 1622
99. Markowitz MA, Bielski R, Regen SL (1988) J. Am. Chem. Soc. 110: 7545
100. Ishikawa Y, Kunitake T, Matsuda T, Otsuka T, Shinkai S: J. Chem. Soc., Chem. Commun. 1989: 736
101. Alberts AH, Cram DJ (1977) J. Am. Chem. Soc. 99: 3380
102. Tabushi I, Kobuke Y, Ando K, Kishimoto M, Ohara E (1980) J. Am. Chem. Soc. 102: 5948
103. Tabushi I, Kobuke Y, Yoshizawa A (1984) J. Am. Chem. Soc., 106: 2481
104. Shinkai S, Koreishi H, Ueda K, Manabe O: J. Chem. Soc., Chem. Commun. 1986: 233
105. Shinkai S, Koreishi H, Ueda K, Arimura T, Manabe O (1987) J. Am. Chem. Soc. 109: 6371
106. Nagasaki T, Arimura T, Shinkai S (to be submitted)
107. Shinkai S, Shirahama Y, Satoh H, Manabe O, Arimura T, Fujimoto K, Matsuda T: J. Chem. Soc., Perkin Trans. II 1989: 1167

Subject Index

MIX
Papier aus verantwortungsvollen Quellen
Paper from responsible sources
FSC® C105338

If you have any concerns about our products,
you can contact us on
ProductSafety@springernature.com

In case Publisher is established outside the EU,
the EU authorized representative is:
Springer Nature Customer Service Center GmbH
Europaplatz 3, 69115 Heidelberg, Germany

Printed by Libri Plureos GmbH
in Hamburg, Germany